Chemistry of the Body

Janet E. Garden

Thornlea Secondary School

Robert A. Richardson

Oakville-Trafalgar High School

John Wiley & Sons

Toronto New York Chichester Brisbane Singapore

The metric usage in this text has been reviewed by the Metric Screening Office of the Canadian General Standards Board. Metric Commission Canada has granted permission for the use of the National Symbol for Metric Conversion.

Cover by Jack Steiner Graphic Design
Illustrations by James Loates

Canadian Cataloguing in Publication Data

Garden, Janet E.
Chemistry of the body

ISBN 0-471-79740-5

1. Molecular biology. I. Richardson, R.A. (Robert A.)
II. Title.

QH506.G37 1985 574.8′8 C85-098207-3

Printed and bound in Canada

10 9 8 7 6 5 4 3

Table of Contents

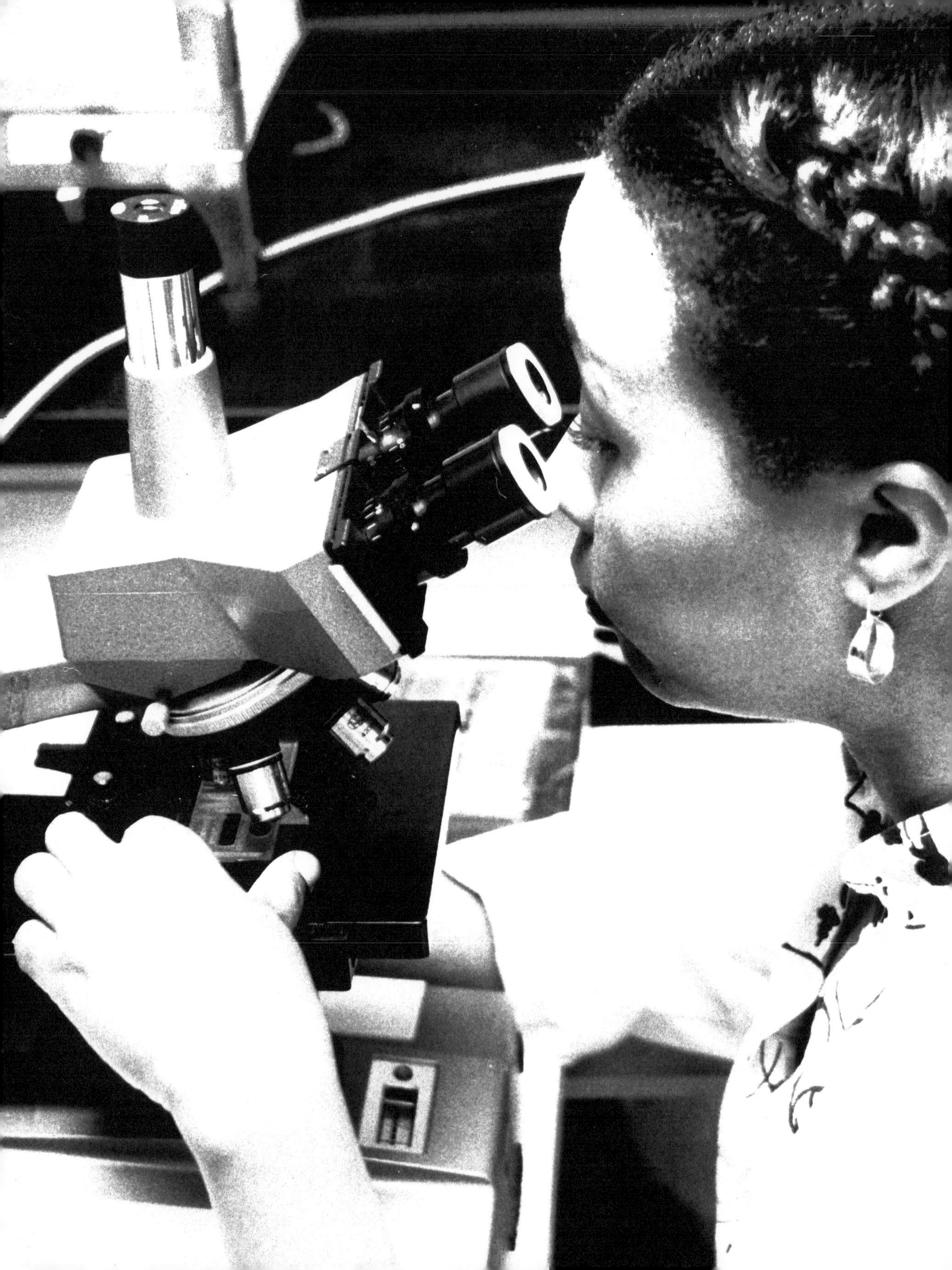

Preface

Biochemistry is one of the most exciting fields in biology today. The rapid development of techniques for the identification and tracing of minute quantities of organic chemicals has made it possible to unravel many complex biological processes. From this advance has come a greater understanding of the elaborate control systems which keep our bodies functioning normally. Knowledge about the structure and functioning of the DNA molecule in particular promises to lead to astonishing capabilities in the prevention and treatment of inherited biochemical defects. It also offers the prospect of synthesizing medical biochemicals that will produce fewer side effects than those presently in use.

This small book has been written in the belief that it is essential for each of us to understand how our bodies function, so that we may protect ourselves from harmful influences. It is also important for us to have the background to realize the implications of biological research, so that we may apply its results to greatest advantage. An informed population is a most effective check on the direction of scientific research, ensuring that it continues to benefit humanity.

Although written primarily as a companion volume to *Biology of Ourselves* by Gordon S. Berry, published by John Wiley & Sons Canada Limited, *Chemistry of the Body* could be used to supplement any standard biology text, since it deals completely with the selected topics.

It will have served its purpose well if it has whetted the appetites of its readers sufficiently that they will keep abreast of the rapidly occurring developments in biological research.

Acknowledgements

We are greatly indebted to Trudy Rising of John Wiley & Sons Canada Limited who perceived the need for this volume and provided the direction and encouragement needed for its development. Others at John Wiley who have been especially helpful in its production are Doug Macnamara, Sonia Skirko, and Kathryn Dean. Special thanks must also be extended to Ed Gillan, Pam Heron, and Pat Wright, who read various portions of the manuscript and provided constructive criticism at times that we are sure were far from convenient. We are also grateful to Bill Kolysher and Peter Moir for their critiques of draft material for the first three sections of this text.

Chapter 1

Atoms and Life

- The Structure of Atoms
- Why Atoms Stick Together
- Covalent Bonding
- Ionic Bonding
- Polar Bonding
- Acids and Bases
- The pH Scale
- Controlling pH
- Energy in Chemical Bonds
- Forming Large Molecules

Chapter 1

Atoms and Life

The living world and the nonliving world have something in common. Both are made of tiny particles called **atoms**. Concrete and grass, bicycles and people are all made of atoms.

The tremendous variety of matter in the universe might lead us to believe that there are millions of different kinds of atoms. In fact, only ninety different kinds of atoms occur in nature. A substance that contains just one kind of atom is called an **element**. Therefore, there are ninety natural elements. Of these, the four elements shown in Figure 1.1 account for most of the structure of living things: carbon (C), hydrogen (H), oxygen (O), and nitrogen (N). Many other elements are important to life, but they are present in very small quantities. Iron (Fe) has an important role in enabling blood cells to carry oxygen. Calcium (Ca) is present in teeth and bones. Phosphorus (P) is also present in teeth and bones and is essential for providing energy for the activities of all living cells. These and a few others are necessary for life, but the "big four" account for most of the body's mass. Ninety-six percent of your body is made of carbon, hydrogen, oxygen, and nitrogen.

Figure 1.1.
The Proportion of Atoms in the Human Body. Hydrogen, carbon and oxygen, found in carbohydrates, fats and proteins, form the major proportion of the body. Nitrogen, needed to form proteins, is also present in significant amounts.

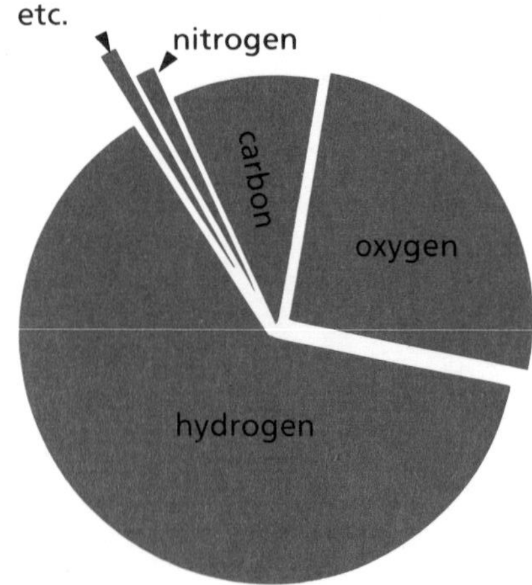

Processes that occur in living things are often similar to or the same as processes in the nonliving world. The breath you exhale is very similar to the gases that rise up the chimney from the furnace in your home. The heat that radiates from your body is produced in a way similar to the heat released in your furnace.

In the furnace and in your body, fuel is being consumed. Atoms are being rearranged to form new substances, and the formation of new substances is accompanied by energy changes. Energy is the ability to do work; therefore, living things need energy for tasks such as feeding, muscle contraction, moving, and growing. You can think of chemical change as the source of the energy that permits your body to function.

The study of body processes, therefore, quickly leads to

a consideration of the chemical reactions involved. In order to understand the chemical reactions of the body, we must first understand why atoms join together and why they can rearrange themselves to form different substances. Although these processes are not fully understood, there are some basic principles that have been found to govern the way atoms attach to each other.

The Structure of Atoms

Atoms are exceedingly small. Two grams of carbon contain approximately 100 000 000 000 000 000 000 000 atoms! Despite their small size, much is known about atoms. Most of their mass is concentrated in a central part called the **nucleus**. The nucleus is made up of particles called **protons** and **neutrons**. These particles are nearly equal in mass, but the proton is much more important in determining how atoms join together to form compounds. Each proton carries a single unit of positive electrical charge. This positive electrical charge is exactly balanced by the negative electrical charge on another kind of particle, called the **electron**. The electron has much less mass than a proton or a neutron. Each electron is located outside of the nucleus, and moves around it very rapidly.

Figure 1.2 shows the arrangement of protons, neutrons, and electrons in a carbon atom. Notice that there are six protons and six electrons. Atoms are electrically neutral when they have the same number of protons and electrons. Figure 1.3 shows the structures of hydrogen, oxygen, and nitrogen atoms. There is no exact rule governing the number of neutrons in an atom. The most important number is the number of protons. This number is the "tag" that identifies the atom. If it has six protons, it is a carbon atom. The presence of seven protons indicates a nitrogen atom.

The arrangement of electrons in the space around the nucleus is complex, and it is very difficult to draw a diagram that is both correct and useful. Some electrons are more energetic than others, and fly high and fast. Others are slower and closer to the nucleus. We can think of electrons as being arranged in **energy levels** around the nucleus. The basic pattern for small atoms is that the first energy level can hold two electrons, the second can hold eight, and the third also can hold eight.

Figure 1.2.
A Carbon Atom. Notice that the number of protons is equal to the number of electrons. (The particles in this atom and in the following figures are not drawn to scale. Electrons are far smaller than protons or neutrons.)

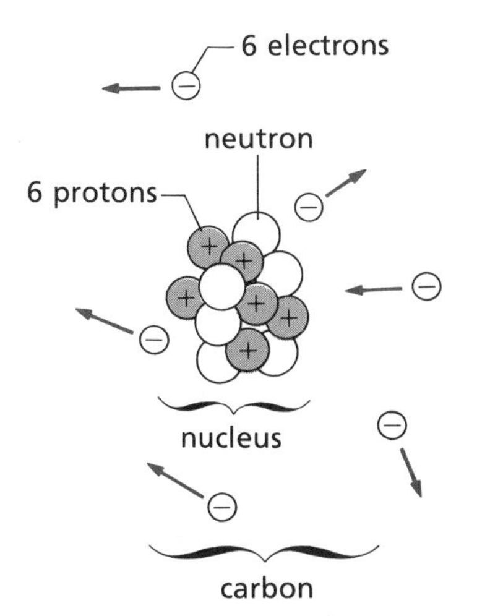

Figure 1.3.
These three elements, along with carbon, make up 90% of human body mass.

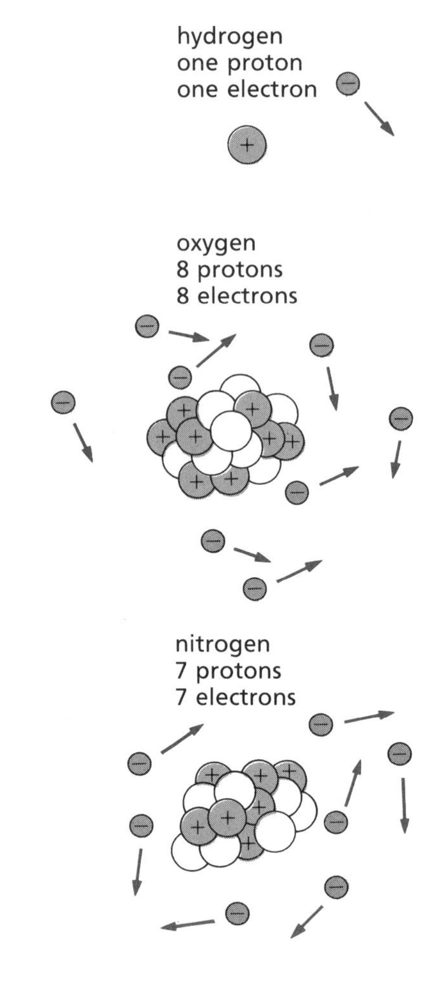

Model of Electrons. This photograph of a beam of light from a rapidly moving flashlight, can serve as a model for the random motion of electrons in the first energy level around the nucleus of an atom.

Many books show electrons arranged in concentric circles. We will use such drawings here, but it is important to realize that electrons do not revolve around the nucleus like planets around the sun. (If you are curious about electrons, and would like to know more, look up "orbitals" in a chemistry textbook.)

Figure 1.4 shows the electron energy levels for the ten lightest atoms. In chemical reactions, the outer energy level is the important one. When two atoms combine, it is the outer electrons which are responsible. Electrons usually occur in pairs, but can remain alone if there is enough space in the energy level. A maximum of four can exist alone before pairing up to a maximum of eight. The arrangement of electrons in the outer energy levels is therefore described in terms of pairs and singles. For example, in the outer energy level, carbon has four single electrons, hydrogen has one single, oxygen has two pairs of electrons and two singles, and nitrogen has one pair and three singles. The positions of the single electrons determine where other atoms can join with that atom. It is possible to predict the shape of the resulting particle, as Figure 1.5 indicates.

Figure 1.4.
Electron Arrangement in the Ten Lightest Atoms

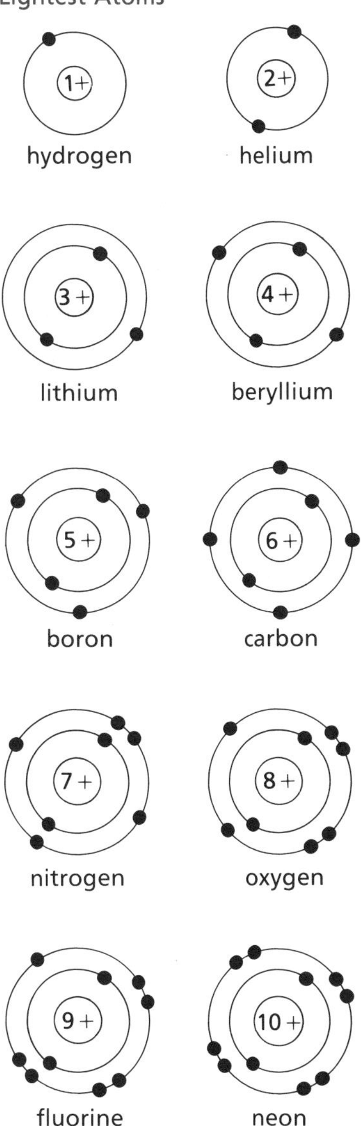

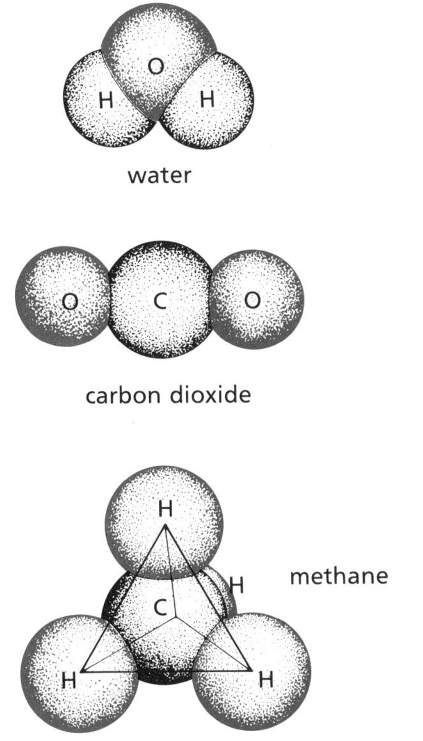

Figure 1.5.
The positions of the single electrons in the outer energy level determine how atoms can fit together. Thus water, carbon dioxide, and methane molecules have very different shapes.

Why Atoms Stick Together

Atoms are electrically neutral when the number of electrons equals the number of protons. However, atoms can gain, lose, or share electrons when they combine. In many cases, the outer energy levels of the combining atoms are completed by this process. Gaining, losing, or sharing electrons always involves energy changes. It is the combination (or break-up) of atoms that enables cells to control the flow of energy and materials.

When two atoms come close to each other, the protons (+) in one atom begin to attract the electrons (−) of the other, and *vice versa*. Figure 1.6 shows what happens when two fast-moving hydrogen atoms bump into an oxygen atom. The single electron of each hydrogen atom pairs with one of the two single electrons in the outer energy level of the oxygen atom. The three atoms share their single electrons with one another. Since it now holds eight electrons, the outer energy level of the oxygen atom is complete. The outer energy level of each of the hydrogen atoms is also complete. (Remember that the first energy level is "full" when it has two electrons, and hydrogen has only one energy level.)

You will recognize H_2O as water, one of the substances essential for our existence. Water is always H_2O; it is never HO_2 or HO_3 or H_4O_7 or any other arrangement. When atoms join together in a definite proportion — such as two hydrogen to one oxygen in water — the new substance is called a **compound**.

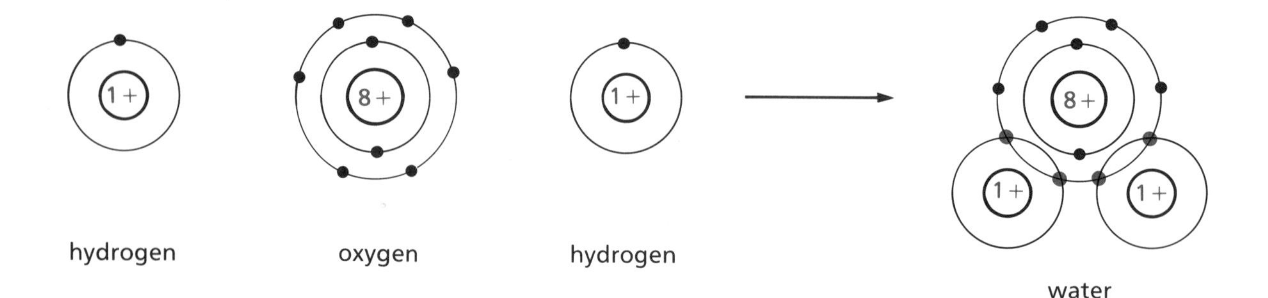

Figure 1.6.
The Bonding of Hydrogen and Oxygen. The hydrogen atoms collide with the oxygen atom, and then share electron pairs with it. All three atoms end up with full outer energy levels. The result is H_2O, water.

Covalent Bonding

If a compound is formed by the sharing of electrons, as in water, it is called a **covalent compound**; the forces holding the different atoms together are called **covalent bonds**. The entire particle — in this case, H_2O — is known as a **molecule**. It is appropriate that the first molecule we have considered is the water molecule. The human body is approximately 60% water. Blood is over 80% water. It is impossible to understand the chemistry of living things without some understanding of the chemistry of water.

The position of the shared pairs of electrons in a covalent compound determines the shape and behaviour of the molecule. For this reason, the composition of these compounds is often shown by a **structural formula**, as shown in Table 1.1. There are two common types of structural diagrams.

Table 1.1
Different Ways of Representing Covalent Compounds. In the electron dot diagrams, only the electrons in the outer energy level are shown. In the structural formulas, shared pairs of electrons are represented by dashes. In the condensed structural formulas, atoms are grouped into the clusters they form in the molecule.

<table>
<tr><th>Molecule</th><th>Electron Dot Diagram</th><th>Structural Formula</th><th>Condensed Structural Formula</th></tr>
<tr><td>Hydrogen gas
H_2</td><td>H : H</td><td>H – H</td><td>H_2</td></tr>
<tr><td>Oxygen gas
O_2</td><td>:O :: O:</td><td>O = O</td><td>O_2</td></tr>
<tr><td>Water
H_2O</td><td>H
:O:
H</td><td>H\
O
H/</td><td>HOH</td></tr>
<tr><td>Carbon dioxide
CO_2</td><td>:O: C :O:</td><td>O = C = O</td><td>CO_2</td></tr>
<tr><td>Methane
CH_4</td><td>H
H : C : H
H</td><td>H
|
H–C–H
|
H</td><td>CH_4</td></tr>
<tr><td>Dimethyl ether
C_2H_6O</td><td>H H
H:C:O:C:H
H H</td><td>H H
| |
H–C–O–C–H
| |
H H</td><td>CH_3OCH_3</td></tr>
<tr><td>Ethanol
C_2H_6O</td><td>H H
H:C:C:O:H
H H</td><td>H H
| |
H–C–C–OH
| |
H H</td><td>CH_3CH_2OH</td></tr>
</table>

In one, only the electrons in the outer layer are shown; they are represented by dots. In the other type, the shared electrons are represented by dashes. The number of dashes will normally equal the number of unpaired electrons in the outer layer. These dashes, therefore, designate **bonding sites** — locations at which atoms can readily join together or split apart during reactions.

The energy level diagrams of Figure 1.4 can be used to predict how atoms will combine to form compounds. From the figure, it is apparent that carbon has four bonding sites. This provides a great number of possibilities for combining with other atoms, even other carbon atoms. Carbon atoms readily link together to form chains, often in more than one direction. The number of possible carbon compounds is immense.

Some Important Carbon Compounds

Molecules which contain only carbon and hydrogen are called **hydrocarbons**. Many fuels, solvents, plastics, and pesticides are hydrocarbons. Carbon-based or *organic* molecules in which the proportion of hydrogen to oxygen is two to one, as in water, are called **carbohydrates. Lipids**, a term that applies to fats and oils, also contain only carbon, hydrogen, and oxygen, but the proportion of oxygen is much lower than in the carbohydrates. **Proteins** are very long chain organic molecules which also contain nitrogen and sometimes sulphur.

The Relationship between the Structure and the Behaviour of Atoms

A structural formula is particularly useful in representing an organic molecule, since it shows the position of each atom. As you will learn later, certain processes depend upon the specific shapes of the molecules involved. The chemical behaviour of a molecule is also determined by the arrangement of the atoms within the molecule. A slight variation in the position of even one atom can change the properties of compounds considerably. It is thus possible for two or more organic molecules to have the same numbers of each type of atom, yet be very different substances.

Consider the last two molecules shown in Table 1.1. Each of these could be represented by the formula C_2H_6O. Yet the different position of the oxygen atom in each molecule produces two substances that have quite different chemical properties. We would not wish to confuse ethanol, the alcohol found in beer and wine, with dimethyl ether, an anesthetic! Structural formulas are useful because it is important to be able to distinguish between **isomers** (the term used to describe chemical compounds with different arrangements of the same component atoms).

Yet structural formulas are clumsy to work with. The solution to the problem has been found in the **condensed structural formulas** shown in the last column of Table 1.1. In this system, the atoms associated with each carbon atom are written as a group. Such formulas are easy to write but the distinction between substances is clear.

Bonding and Fluids

The compounds illustrated in Table 1.1 are all **fluids**. That is, they are all liquids or gases, and can flow. Liquids and gases are important in the chemistry of life. Blood, urine, and lymph are water with many other substances dissolved or suspended in it. The air we inhale is mostly nitrogen and oxygen with traces of other gases. The breath we exhale contains less oxygen and more carbon dioxide. Can we account for the behaviour of fluids by considering their bonding?

Examine the diagram of the water molecule in Figure 1.6. It shows how electrical forces hold the two hydrogen atoms near the oxygen atom, but does not suggest any obvious way for one water molecule to attach itself to another water molecule. All the single electrons in the outer energy levels of the hydrogen and oxygen atoms form bonds within the molecule. There are no bonding sites left over to provide connections to neighbouring molecules. Therefore, each water molecule can slide past its neighbours very easily. In fact, water molecules do stick to each other slightly. If they slow down enough (as a result of cooling, for instance), the sticking becomes pronounced and ice is formed. (Later on, we will have to revise our concept of water molecules to account for forces between them. But at room temperature, these forces are not very strong, and water flows.)

There are many substances which flow even more easily than water, among them, methane and oxygen. At room temperature, these are gases. In general, substances formed by covalent bonding flow, and are gases, liquids, or soft solids at room temperature. However, some covalent compounds contain long chains and networks of atoms, and may be solids that are quite tough and do not flow at all. The protein that forms cartilage is an example of a covalently bonded substance that is extremely strong.

Ionic Bonding

There is another category of substances, in which the bonding is quite different from covalent bonding, but equally important. Consider table salt. The chemical name for salt is sodium chloride, and its formula is NaCl. Salt is hard and brittle, and formed of tiny cube-shaped crystals.

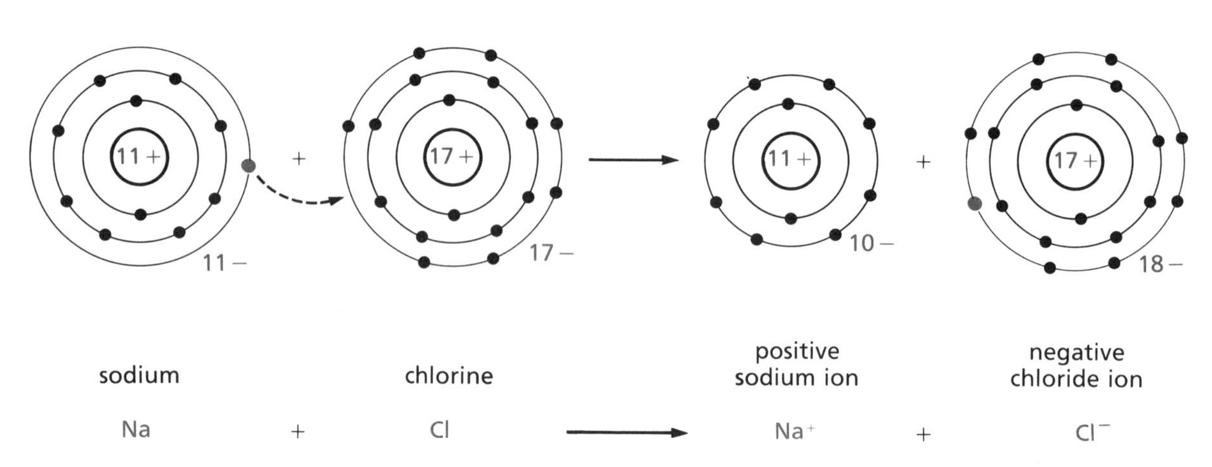

Figure 1.7.
The Formation of Sodium Chloride, NaCl. Sodium loses an electron to chlorine to form two oppositely charged ions. These are then attracted to each other to form the ionic compound, table salt.

Figure 1.7 shows the formation of sodium chloride. Notice that the sodium and chlorine atoms both attain full energy levels, but in a different way from covalent bonding. An electron actually moves from the sodium atom to the chlorine atom. Before this electron transfer, the sodium atom has 11 protons and 11 electrons. The chlorine atom has 17 protons and 17 electrons. Both atoms are electrically neutral. After the electron moves from the sodium to the chlorine, the sodium atom still has 11 protons, but only 10 electrons. It has one less electron than it has protons; therefore, it has a positive electrical charge. The chlorine atom still has 17 protons, but now it has 18 electrons; that is, it has one more electron than it has protons. It therefore has a negative electrical charge. As you know, one of the basic laws of electricity is that objects with opposite charges attract each other. Thus, the positively charged sodium is strongly attracted to the negatively charged chlorine. Atoms that have become charged by loss or gain of electrons are called **ions**. Compounds made up of positive and negative ions are called **ionic compounds**. In these compounds, the positive ion keeps its name but the negative ion's name ends in "-ide", as in sodium chloride.

Figure 1.8 shows the arrangement of ions in sodium chloride. Each positive ion has six negative ions as nearest neighbours, and *vice versa*. Ionic compounds are very tightly held together by the attractions between their positive and negative ions; therefore, all ionic compounds are solids at room temperature. They do not flow, but many of them dissolve in water to play important roles in the chemistry of living things.

Figure 1.8.
A Model of Sodium Chloride. Each positively charged ion is surrounded by six negatively charged ions (left, right, front, back, above, and below). Similarly, each negative ion is surrounded by six positive ions. Electrical attractions make this a strong structure.

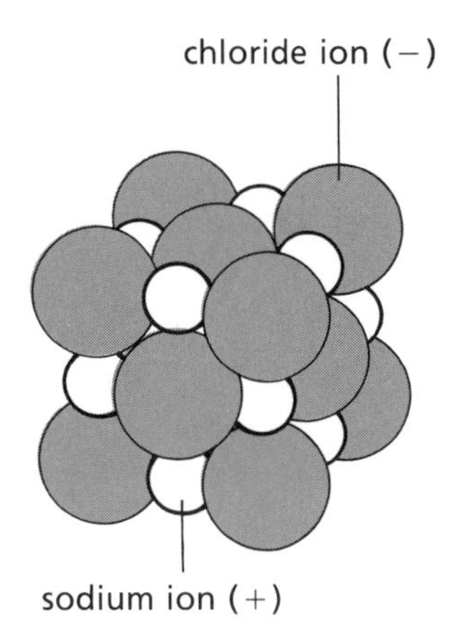

Dissolving Ionic Compounds

When ionic compounds dissolve in water, they *dissociate* into their component ions. This makes the ions readily available for reaction, yet because they are close together and present in the correct proportion, the properties remain those of the ionic compound, not those of the separate elements. Thus, when you put salt on your food, it dissolves in the water in your stomach. Fortunately, however, the sodium and chloride ions still behave as the ionic compound, table salt, not as the elements sodium and chlorine, both of which would poison you.

Some elements tend to cluster together as a group which then acts as an ion. When they are in solution, these elements remain together as a unit. Some examples of these clusters are given below:

HCO_3^-	hydrogen carbonate	NO_2^-	nitrite
CO_3^{2-}	carbonate	NO_3^-	nitrate
ClO_3^-	chlorate	PO_4^{3-}	phosphate
OH^-	hydroxide	SO_4^{2-}	sulphate

Consider baking soda, sodium hydrogen carbonate ($NaHCO_3$), as an example. When it is dissolved in water, the ions which result are Na^+ and HCO_3^-. These clusters of atoms, which behave as ions, are very important to the maintenance of the correct acid/base balance in the body.

Polar Bonding

Water has been called "the universal solvent". Its ability to act as a solvent makes it a good transport medium in the body. Why do so many substances dissolve in water?

You saw in Figure 1.6 that electrons are shared in a water molecule, not transferred from one atom to another. Suppose for a moment that they *were* transferred. How would water behave? It would behave like salt! If water were an ionic compound, it would be solid at room temperature.

Although water is *not* an ionic compound, there is evidence that water molecules stick to each other somewhat, even in the liquid state. The cohesive forces between water molecules are much weaker than the forces in an ionic bond or a covalent bond, but they are strong enough to give water some interesting properties.

Perhaps you have noticed that a steel needle can be made to float on water if it is placed on the surface very carefully. Similarly, insects called water striders can walk on water in a manner similar to a gymnast walking on a trampoline. These observations suggest that there is a kind of elastic surface on water. We call this phenomenon **surface tension**. Surface tension provides evidence that water molecules do indeed attract each other in some way.

By observing thousands of compounds containing oxygen, scientists have discovered that oxygen has a very strong attraction for electrons. The ability of an atom to attract the electrons in a bond is called its **electronegativity**. Electronegativity is generally greater in small atoms and in atoms with close to eight electrons in the outer energy level. Oxygen is one of the smallest atoms, and it has six electrons in its outer energy level. Oxygen, therefore, is a highly electronegative element, and oxygen atoms that are bonded to other kinds of atoms will exert a strong pull on the electron pairs in the bond. For example, the shared pairs of electrons in a water molecule spend more time near the oxygen atom than near the hydrogen atoms. In other words, the sharing is unequal. Since electrons have negative charges, this inequality means that the oxygen atom in a water molecule is a little bit negative and the hydrogen atoms are a little bit positive. In Figure 1.9, these small charges are shown by the small Greek letter *delta* (δ). Figure 1.10 shows how the small charges cause water molecules to attract each other.

Molecules that have charges caused by unequal sharing of electrons in covalent bonds are called **polar** molecules. As well as causing water molecules to attract each other, these small charges cause them to attract other charged objects, such as ions. This fact explains why ionic substances like salt dissolve readily in water. Figure 1.11 shows how water molecules attract the ions in a salt crystal, and cause the salt to dissolve. Many elements are found in body fluids as ions. Iron, magnesium, and calcium ions, for example, are important in blood chemistry. Sodium and potassium ions play a role in transmitting electrical signals through the nerve fibres. It is the function of the kidneys to maintain a proper balance of these ions in body fluids.

Figure 1.9.
Charges on the Water Molecule. Unequal sharing of the electrons in the bonds makes the oxygen atom slightly negative and the hydrogen atoms slightly positive.

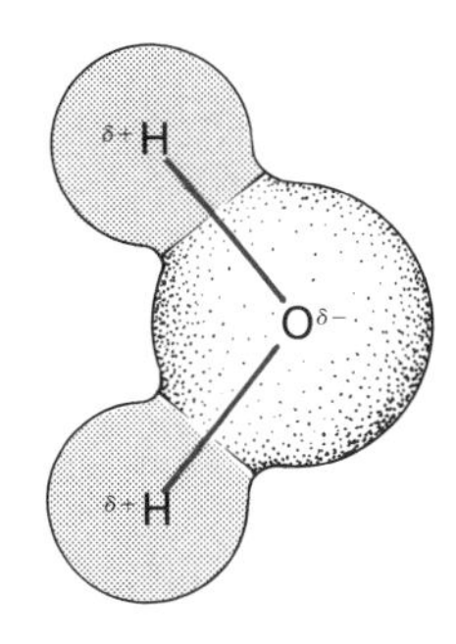

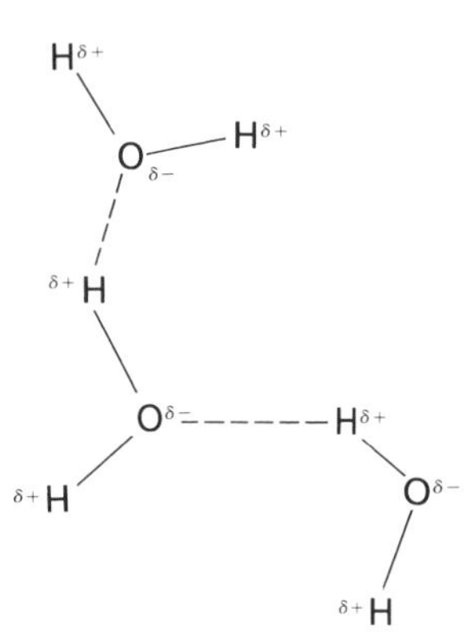

Figure 1.10.
Attractive Forces between Water Molecules. The negative part of one molecule attracts the positive part of a neighbour. The attractive forces are shown by dotted lines.

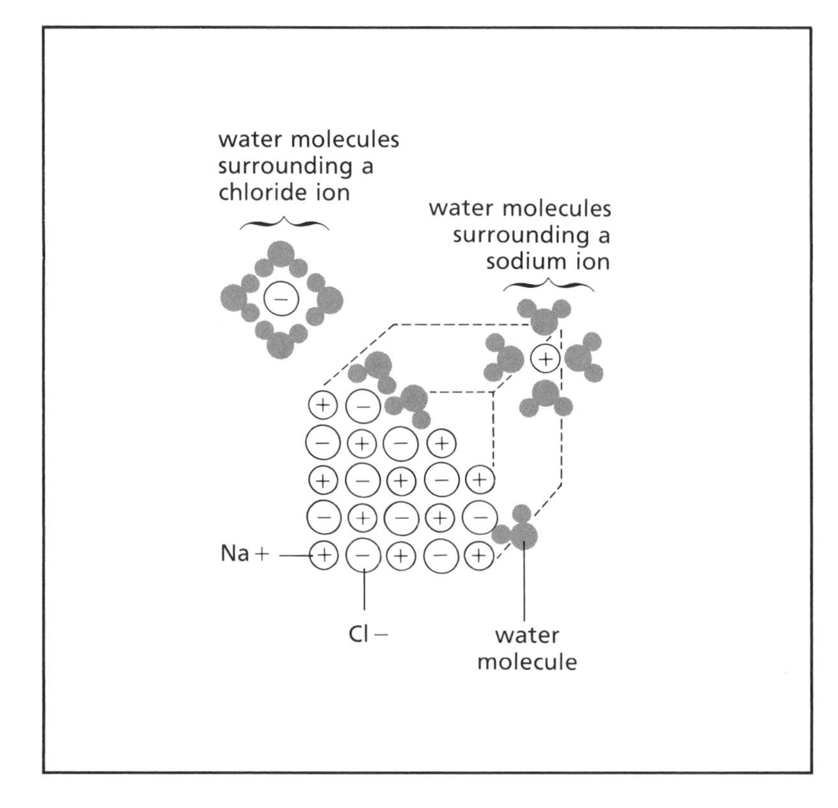

Figure 1.11.
How Salt Dissolves. The polar water molecules attach themselves to the ions in the crystal. The more positive (H) portions of the water molecule are attracted to the chloride ion, and the more negative (O) portion, to the sodium ion. Surrounded by water molecules, the ions drift away into the solution.

Acids and Bases

When some substances ionize in water, the solution may become either *acidic* or *basic*. Both acids and bases tend to be corrosive and must be handled with caution, yet their properties are so different that they can be considered as opposites. Not all acids and bases are dangerous — vinegar and lemon juice are acids; baking soda is a base.

Acids and bases can easily be distinguished by their effect on **chemical indicators**, substances which rearrange their atoms when they are exposed to acids or bases. When this rearrangement happens, an observable colour change is produced. A very sensitive chemical indicator used frequently in biology is bromothymol blue. This substance is blue in a basic solution, green in a neutral one, and yellow in an acidic solution.

Much of our food is somewhat acidic. Pickles, sauerkraut, and lemonade come to mind. The cells lining the stomach produce large quantities of hydrochloric acid to aid in digestion. Many of the waste products of metabolism are also acidic. These acids would destroy cells if they were allowed to accumulate. To understand how the body solves this problem, you need to understand how acids and bases behave.

The Behaviour of Acids and Bases

To account for the behaviour of acids and bases, we have to think again about the lightest element, hydrogen. A hydrogen atom is made up of a single proton and a single electron. If the electron is removed, the atom is turned into a hydrogen ion (H^+). This is a very special ion. The positive charge is concentrated in a very small mass — just one proton. As a result, the hydrogen ion is very attractive to other particles. Hydrogen ions are seldom found alone; rather, they tend to participate in chemical reactions. Changes in which a hydrogen ion is lost or gained are called **proton transfers**. In proton transfers, the particle that loses a proton is called an acid, while the particle that gains the proton is called a base.

A very important acid found in living things is hydrochloric acid, HCl. When an HCl molecule enters water, it loses its proton to a water molecule. In this case, the HCl molecule is the acid and the water molecule is the base.

This event can be represented pictorially as in Figure 1.12, or written in the shorthand of a chemical equation:

$$H_2O + HCl \longrightarrow H_3O^+ + Cl^-$$

Figure 1.12.
The reaction of an Acid with Water. The water molecule attracts the hydrogen atom from HCl. But the hydrogen atom leaves its electron behind. The H_3O^+ ion formed is called the hydronium ion. The Cl^- is the chloride ion.

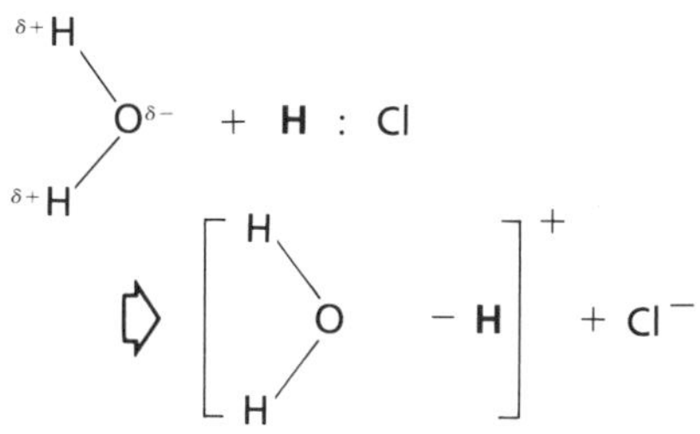

The ion H_3O^+, called the **hydronium ion**, is present in the water solutions of all acids. Think of it as a captured proton attached to the water molecule.

Sometimes the water molecule is the acid rather than the base. When ammonia, NH_3, is dissolved in water, some of the ammonia molecules take a proton from the water molecules:

$$NH_3 + H_2O \longrightarrow NH_4^+ + OH^-$$

Ammonia is a typical base in that it causes OH^- ions to form in water. The ion OH^-, called the **hydroxide ion**, is present in the water solutions of all bases.

There are so many natural acids and bases in our environment that hydronium and hydroxide ions are present in all water solutions. Indeed, even pure water forms hydronium and hydroxide ions all by itself. Approximately one water molecule in each 500 million splits apart to form a hydronium ion and a hydroxide ion.

The pH Scale

A substance such as hydrochloric acid that produces a high concentration of hydronium ions in water is called a *strong acid*. A substance such as sodium hydroxide that produces a high concentration of hydroxide ions in water is a *strong base*. (You should take care not to confuse "strong" in this context with "concentrated". Some acids such as acetic acid (vinegar is dilute acetic acid) don't donate their protons easily, so it is possible to have a concentrated solution that is not strong!)

The acid/base character of a solution is often expressed on a 0–14 scale called the **pH scale**. The mid-point of the scale, 7, has special significance. This is the pH of a sample of pure water or any water solution in which the hydronium and hydroxide concentrations are equal. If the hydronium ions outnumber the hydroxide ions, as they do in an acid

Figure 1.13.
The pH Values of Some Common Substances

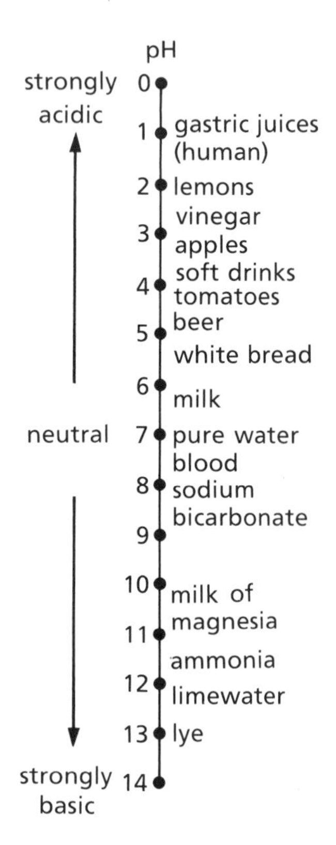

solution, the pH is less than 7. If the hydroxide ions are more numerous than the hydronium ions, as in a basic solution, the pH is greater than 7.

A change of one pH unit represents a change by a factor of 10 in the hydronium and hydroxide ion concentrations. For example, if you add acid to a solution of pH 8 to bring the pH down to 7, the hydronium ion concentration increases by a factor of 10 and the hydroxide ion concentration decreases by a factor of 10. Figure 1.13 shows the pH of some common substances.

Acids and Bases Together: Neutralization

Using a base to *neutralize* an acid — or *vice versa* — can be done easily in the laboratory. Strong drain cleaners contain the very dangerous base sodium hydroxide. When sodium hydroxide dissolves in water, it produces hydroxide ions.

$$NaOH \longrightarrow Na^+ + OH^-$$

Hydroxide ions have a strong affinity for protons. So, when hydrochloric acid is mixed with a sodium hydroxide solution, the acid loses its proton and the hydroxide gains the proton to become water:

$$HCl + OH^- \longrightarrow H_2O + Cl^-$$

or

$$HCl + NaOH \longrightarrow H_2O + NaCl$$

In this example, two dangerous substances, through neutralization, have turned into edible substances, table salt and water!

The above process is an interesting example of the way a chemical reaction can produce a striking change in matter. However, this example should not be tried outside the laboratory. Sodium hydroxide and hydrochloric acid are equally dangerous. There is no point in pouring hydrochloric acid on spilled drain cleaner when the hydrochloric acid is itself dangerous. The normal procedure is to treat a base spill with a weak acid, and an acid spill with a weak base. In either case, lots of water can be used to dilute the spill if specific remedies are not available.

Water is capable of functioning as either an acid or a base. Since water is only slightly ionized, generous quantities must be used. Water has several advantages in such a situation:

you need not know if the spilled substance is an acid or a base; water (or a substance that is mostly water) is readily available; water can absorb the heat released during neutralization reactions. Only water should be used if the acid or base has been swallowed or spilled on the skin.

Controlling pH

Many of the chemical reactions of the body involve acids and bases. Yet the maintenance of suitable pH levels in the body fluids is essential for health. This is accomplished through a remarkable process called **buffering**.

During digestion, the pH of the stomach contents is held between 1.2 and 3. Yet blood pH is very closely regulated at pH 7.40. It is remarkable that blood pH remains this constant when the foods and medicines we consume range from about pH 2 for lemon juice to around pH 10 for milk of magnesia. How is regulation achieved when our intake varies so much?

The most important buffering system in body fluids is provided by carbonic acid and its close relative, the hydrogen carbonate ion. (Another name for the hydrogen carbonate ion is the "bicarbonate ion".) Carbonic acid forms in water solutions, such as blood, when carbon dioxide dissolves in it.

$$H_2O + CO_2 \longrightarrow H_2CO_3$$

When carbonic acid loses one of its protons (H^+), it forms the hydrogen carbonate ion (HCO_3^-)

$$H_2CO_3 \longrightarrow H^+ + HCO_3^-$$

If the pH of the blood should fall below its normal level, that is, if the blood should become too acidic, some of the hydrogen carbonate ions will capture excess protons. The excess acid is thus neutralized, and the pH returns to normal.

$$HCO_3^- + H^+ \longrightarrow H_2CO_3$$

If blood pH should rise above its normal level, that is, if the blood should become too basic, some carbonic acid molecules will release protons. The excess base is thus neutralized, and the pH again returns to normal.

$$OH^- + H_2CO_3 \longrightarrow HCO_3^- + H_2O$$

In the digestive system, too, control of pH depends on buffers. The pancreas releases hydrogen carbonate ions into the first part of the small intestine. This release creates a basic environment suitable for the action of certain digestive enzymes on food.

Energy in Chemical Bonds

Imagine two atoms bonded together like two spheres joined by an elastic band or a spring. It takes energy to pull the spheres further apart. If the spheres are allowed to snap back, energy is released. The amount of energy used to separate the spheres is exactly the same as the amount released when the spheres are allowed to snap back. The physical laws that govern forces in springs and elastics are similar to the physical laws that govern forces in molecules.

Energy is always required to stretch, and then break, a chemical bond between two atoms. Energy is always given off when bonds form. In every cell of every living thing, energy is stored and released by the breaking and forming of chemical bonds. In the complex organization of living things, energy management is of utmost importance.

The amount of energy released when a chemical bond forms is called the **bond energy**. It will be helpful to look at some numbers to gain an understanding of bond energy. The amount of energy released when just one bond forms is very small, much too small to measure. For this reason, biochemists think in terms of a very large number: the **mole**. Just as bakers count doughnuts or rolls in dozens, so biochemists count atoms and molecules in moles. Since the objects being counted are very small and numerous, the mole has to be a very large number indeed. It is equal to 6.02×10^{23}, yet even this large number of water molecules would not fill a test tube.

Scientists are able to measure the mass of this amount of a substance, however, by using a chemical balance of the type you have at your school.

Breaking and Forming Bonds

You have seen that chemical change involves the making and breaking of bonds. Since it is not possible to measure this energy accurately for a single bond, it is the energy required to form or break a mole of bonds that has been determined. Figuring out the overall energy change is something like making entries in a cheque book. Energy is required to break the old bonds, but energy is produced in forming the new ones. The following equation, using structural formulas, shows diagrammatically how bonds between different atoms change during the reaction in which methane is burned.

$$\begin{array}{c} H \\ | \\ H{-}C{-}H \\ | \\ H \end{array} + O{=}O + O{=}O \longrightarrow O{=}C{=}O + \begin{array}{l} H \\ \;\;\diagdown \\ \;\;\;\;O \\ \;\;\diagup \\ H \end{array} + \begin{array}{l} H \\ \;\;\diagdown \\ \;\;\;\;O \\ \;\;\diagup \\ H \end{array}$$

This equation reads, "One molecule of methane reacts with two molecules of oxygen to produce one molecule of carbon dioxide plus two molecules of water." During this reaction, four C–H bonds and two O = O bonds are broken, while two C = O bonds and four O–H bonds are formed. By referring to Table 1.2, which shows the bond energies for some common bonds in living matter, we can calculate the energy given off in the reaction.

Table 1.2 gives the bond energies for some common bonds in living matter. The numbers in Table 1.3 (on the next page) represent the amount of energy used in bond breaking, as well as the amount of energy produced in bond forming.

You will find that 648 kJ more energy is produced than

Table 1.2
Some Average Energies

BOND TYPE	AVERAGE BOND ENERGY (measured in kilojoules per mole)
C – C	344
C = C	610
C ≡ C	835
C – H	415
O – H	463
C – N	305
C – O	350
C = O	725
N ≡ N	949
H – H	436
O = O	497

Table 1.3
Calculating the Bond Energies of a Reaction

ENERGY OF BOND BREAKING	ENERGY OF BOND FORMING
C–H: 4 × 415 = 1660 kJ	C=O:2 × 725 = 1450 kJ
O=O: 2 × 497 = 994 kJ	O–H: 4 × 463 = 1852 kJ
Total = 2654 kJ	Total = 3302 kJ
Net Energy Released = 3302 − 2654 = 648 kJ	

is used when one mole of methane molecules burns. We put this fact to practical use by burning natural gas — which is mostly methane — in our furnaces. Methane burning is an example of an **exothermic reaction**, a reaction which produces energy.

H H
| |
H—C—C—H ethane
| |
H H

a saturated hydrocarbon

H H
| |
C = C ethene
| |
H H

an unsaturated hydrocarbon

H—C ≡ C—H acetylene

an unsaturated hydrocarbon

Figure 1.14.
Three Two-carbon Compounds. The first has a single bond between its two carbon atoms; the second, a double bond; and the third, a triple bond.

Multiple Bonds and Their Energies

Figure 1.14 gives the structural formulas for three simple molecules with single, double, and triple bonds. Notice that ethene contains a double bond, which contains two pairs of electrons. Acetylene has a triple bond, containing three pairs of electrons. Compounds with multiple bonds are said to be **unsaturated**, because the multiple bond uses up more bonding sites, leaving room for fewer hydrogen atoms. Multiple bonds are stronger than single bonds. The explanation for this fact is that a bond is caused by the attraction of pairs of electrons for the nuclei of two atoms. When there are more pairs of electrons, as in multiple bonds, there is a greater force of attraction.

The triple bond in acetylene needs a large quantity of energy to break it, much larger than for the double bond in ethene or the single bond in ethane. Also, acetylene has fewer hydrogen atoms to form H–O bonds, so less energy will be produced in new bond formation. Imagine burning these compounds. In the process of burning, it will be necessary to break some bonds and form new bonds with oxygen atoms. Overall, ethane produces the most energy; acetylene produces the least.

In general, saturated substances — substances that contain single bonds only — produce more energy when they are burned than do similar substances with double or triple bonds (unsaturated substances).

The energy released in the body by food depends, in a similar way, on the type of carbon–carbon bonds in the food molecules. This information is of particular value to those who are trying to control their weight. The energy released from food being used by the cells comes from these chemical bonds. Since unused energy is stored by the body, reducing energy intake can result in weight loss. Fats and oils are both used in the preparation and serving of other foods, yet they are very high in energy themselves. The use of an oil (liquid at room temperature), which contains largely unsaturated bonds, in preference to a fat (solid at room temperature), which contains mostly saturated bonds, can make a considerable difference in weight gain or loss over a period of time.

The Role of ADP and ATP

The cell must do something with the energy it obtains from food. If the energy is not immediately needed, it can be stored in a pair of specialized molecules, adenosine diphosphate (ADP) and adenosine triphosphate (ATP). These molecules cause energy to be stored, just as energy can be stored in a rechargeable battery.

The temporary storage is accomplished by means of phosphate groups (PO_4^{3-}), which are added to ADP to form ATP, or removed from ATP to form ADP. The former process is endothermic; the latter is exothermic, and provides energy when it is demanded by the cell.

How do we know this bond-breaking process is exothermic? The breaking of the O–P bond in ATP requires energy; however, other bonds are being formed as the broken ends are "tied off" by H atoms and OH groups. Bond forming is always exothermic. A careful summing up of all the inputs and outputs shows that the process is a net energy producer, providing body cells with the energy needed for their functions.

You might think that all the ATP in a cell would eventually be used up and the organism would die. The ATP is constantly regenerated. The regeneration process requires energy, and several processes exist for providing the energy. During cellular respiration, for example, glucose is broken down, releasing water, carbon dioxide, and energy. Some of that energy is used to convert ADP to ATP. The cell's energy storage system is thus constantly being recharged, and there is usually enough ATP available to meet the next demand.

The Rate of Reactions

As you have seen, it is possible to predict the amount of energy produced in simple reactions. Tables of bond energies make it possible to add up the inputs and outputs and come up with a net energy change. But such exercises leave a very important question unanswered. How do we know a reaction happens at all? And if it does happen, how rapidly does it proceed? Once again, we have to rely on experimental evidence.

If we set fire to a peanut, it burns fiercely, releasing energy in the form of heat. But if we eat a peanut, its breakdown is, fortunately, much slower. Nevertheless, the same amount of energy is released, and the same products result. Both reactions are exothermic. How does the body control exothermic reactions so its cells are not burned in the process? What is the mechanism that controls the rates of reactions?

Before atoms can rearrange themselves, a certain amount of energy has to be added. This added energy is called the **activation energy**. In the case of the peanut, the activation energy is provided by the chemicals on the end of the match that you apply. Figure 1.15 shows the energy changes during the burning of the peanut. The potential energy of the system is relatively high at the beginning and lower at the end. However, to move from high to low, it must first go

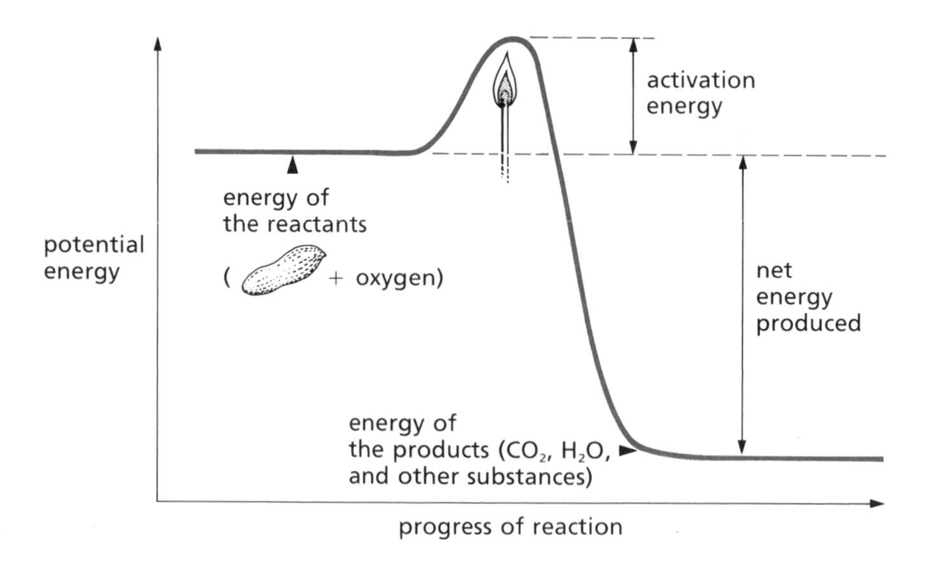

Figure 1.15.
Energy Changes during the Burning of a Peanut. The match supplies the activation energy needed to push the reaction over the energy barrier.

even higher. There is an **energy barrier** that must be overcome. One way to get over it is to heat the reactant(s): we apply a match to the peanut to provide the activation energy. The match flame gives enough energy to the reactant molecules to start the reaction. From then on, the reaction itself produces enough energy to keep going.

But there is another way to surmount the activation energy barrier. It is possible to lower the barrier by the addition of a substance called a **catalyst**. A catalyst provides an alternate pathway for the reaction, as Figure 1.16 shows. It thereby reduces the amount of energy required to initiate the reaction. The reaction proceeds by several steps, one after another. In the end, the same final products are formed and the catalyst emerges unchanged. The important thing about the alternate pathway is that the energy barriers in the new path are lower than the single barrier in the old path. In this way, many reactions which normally occur at high temperatures can be made to occur at body temperature.

Enzymes

Organic catalysts which operate in the body are called **enzymes**. Nearly every chemical process that occurs in living things is made possible by a series of specific enzymes.

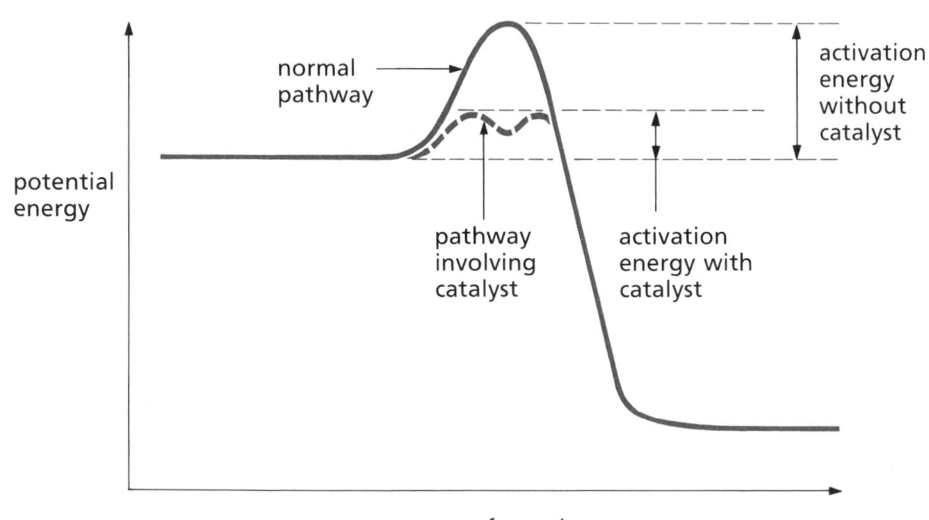

Figure 1.16.
The Effect of a Catalyst. The catalyst provides an alternative pathway. The activation energies in the alternate pathway are lower, so the reactant molecules do not need to be as energetic. The reaction therefore happens at a lower temperature.

Unlike the catalysts you may have met in the chemistry laboratory, enzymes are complex proteins which can catalyze only a specific reaction. They are, therefore, named by the reaction they affect. All enzymes are given the suffix *-ase*. The first part of the name identifies the regulated process. Lactase, for example, controls the conversion of the milk sugar, lactose, into simpler sugars for digestion. Lipase regulates the digestion of lipids.

Compounds upon which an enzyme acts are called **substrates**. Substrates A and B must come together to form

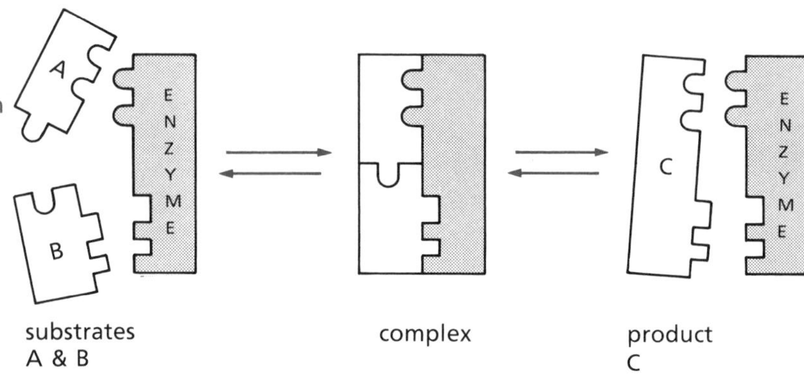

Figure 1.17.
The Lock-and-Key Theory of Enzyme Action. In the reaction A + B → C, substrates A and B join together more efficiently with the help of the specific enzyme.

product C. In the absence of an enzyme, this process could only happen at a very high temperature. Fortunately, as you can see in Figure 1.17, there is an enzyme whose shape fits the shapes of the substrates.

The special shape of each enzyme has prompted the description, "lock-and-key". The lock-and-key theory accounts for the fact that many biochemical reactions require a specific enzyme. For instance, some people are unable to digest most milk products because they lack the one enzyme, lactase, required to break down lactose. These people can usually eat yoghurt, however, because the bacteria used to culture the yoghurt break lactose down into simpler sugars.

When an enzyme and its substrates come close together, weak bonding forces between the matching surfaces hold them together. Once the substrates are in position, the effect of these bonding forces appears to be to distort the substrate molecules sufficiently that the activation energy is lowered. Thus, product C can be formed at the relatively low temperatures that exist in cells. Product C then detaches itself from the enzyme, leaving it available to be reused. These processes occur so rapidly that a single enzyme molecule can catalyze several million reactions a minute! A cell therefore requires only relatively few molecules of any given enzyme.

Factors Influencing Enzyme Action

If an enzyme can catalyze a reaction this rapidly, the cell must have some method of controlling enzyme function. The controlling arrangement is quite intriguing. Enzymes usually function in a series, rather like workers along an assembly line. The end product of the manufacturing process often inhibits the action of the enzyme that starts it. This process is called **feedback inhibition**. In other cases, the end product prevents the very formation of the enzymes — an effect called **enzyme repression**. Thus, only when the product has been consumed by the cell can the enzyme start the process of manufacturing more.

The action of an enzyme may also be blocked if the incorrect substrate fits onto it. The enzyme then finds it difficult or impossible to detach from the intruding substrate. This effect is called **competitive inhibition**. A number of pesticides function in this way, and thereby prevent some normal body process from occurring.

Many enzymes require the presence of an additional substance before they can function. Some enzymes require a **co-factor** — usually a metallic ion. This explains why your diet must contain minute traces of such metals as copper, zinc, iron, and magnesium. The activity of these enzymes can be blocked by the presence of such pollutants as mercury and lead. Other enzymes can only function in the presence of a **co-enzyme**. Co-enzymes are small organic molecules (usually not proteins). Many vitamins function as co-enzymes. Some co-enzymes are firmly attached to the enzyme itself; others are only briefly associated with it.

The action of enzymes is strongly influenced by temperature and pH, as you can see in Figures 1.18 and 1.19 (on the next page). The rate of reactions regulated by body enzymes increases up to about 37°C, body temperature. Above this temperature, the rate is reduced sharply. The most suitable pH for cellular enzymes, as you might expect, is around 7. The enzymes of the stomach function best at a pH of 1 to 2, whereas those of the small intestine function best around pH 8.

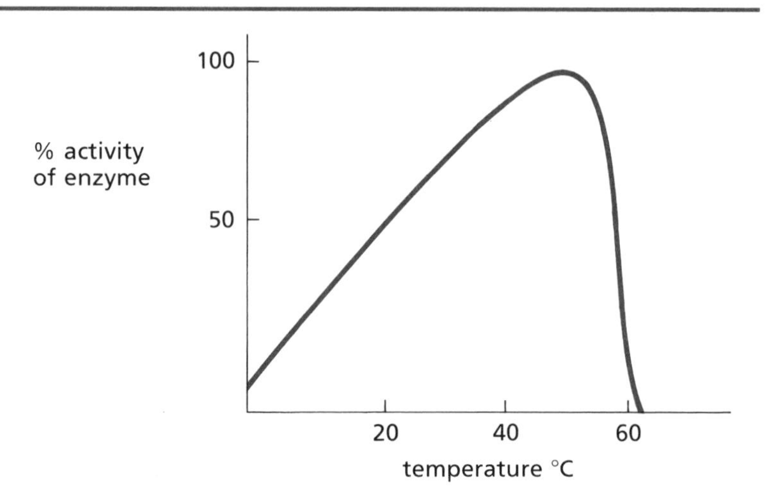

Figure 1.18.
An increase in temperature increases the rate of reaction, including those regulated by enzymes. The maximum activity of enzymes occurs at temperatures just above normal body temperature. The activity decreases rapidly beyond this point, since the structure of proteins, including enzymes, changes at high temperatures.

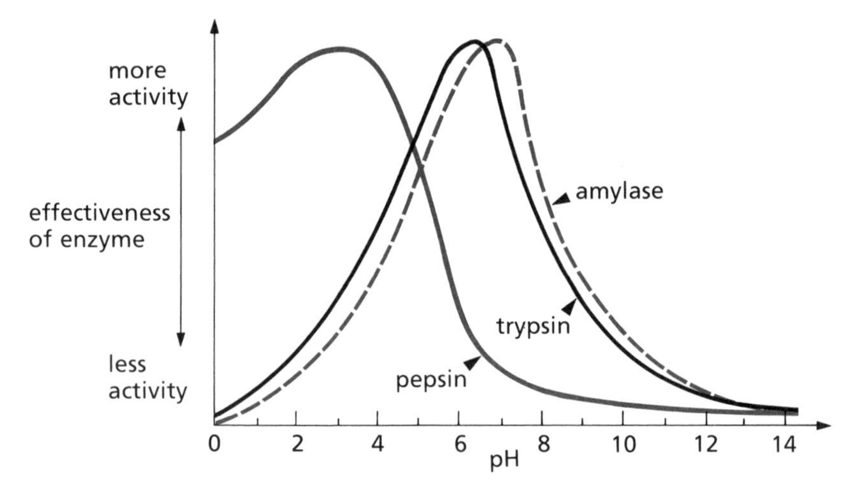

Figure 1.19.
The activity of enzymes depends on factors such as pH. Amylase (found in the saliva and in the small intestine) starts the breakdown of starch. Pepsin works in the acidic conditions of the stomach to start protein digestion. Trypsin works in the small intestine and also breaks down proteins, but works at a higher pH than pepsin. (Note that neither pepsin nor trypsin has the -ase ending typical of enzymes. They were named before the convention for naming came into use. In fact, pepsin was the first enzyme named, in 1825.)

Forming Large Molecules

Large carbon molecules, called **polymers**, are formed by linking simpler molecules together somewhat like beads on a chain. The **synthesis** (building) or **digestion** (breaking apart) of these molecules is usually accomplished by the removal or addition of a molecule of water.

When two simple molecules (sometimes called **monomers**) come together, a hydrogen atom is removed from one and a hydroxyl group is removed from the other. The two molecules then link across the broken bonds. The hydrogen atom and the hydroxyl group come together to form a molecule of water. Because water is removed from the molecules during this process, it is called **dehydration synthesis.**

During digestion, polymers are split apart into their component molecules at these same bonds. Water is required during the process to supply the hydrogen atoms and hydroxyl groups necessary to complete the broken bonds. Since water is added during this splitting process, it is called **hydrolysis**.

These two processes are closely regulated by enzymes and are responsible for the digestion and synthesis of such molecules as carbohydrates, fats, and proteins.

Carbohydrates

Carbohydrates are polymers of simpler molecules called **sugars**. Sugars are named so that they end with *-ose*, as, for example, in sucrose (table sugar). Although many sugars occur in nature, only two groups are important in the body. One group consists of the five-carbon sugars (pentoses). An extremely important pentose is ribose; it is one of the building blocks of DNA and RNA, as you will learn in Chapter 4.

The six-carbon sugars (hexoses) are the ones which cells can use to supply energy. Examples of hexoses are glucose, fructose, and galatose. Since these can all be represented by the formula $C_6H_{12}O_6$, they are isomers of each other. (The structure of glucose is shown in Figure 1.20.) These single sugars are called **monosaccharides**. When two monosaccharides are linked together by dehydration synthesis, as shown in Figure 1.21 (on the next page), the resulting sugar is called a **disaccharide**. Sucrose, lactose, and maltose are examples of disaccharides. Note that they all have the formula $C_{12}H_{22}O_{11}$, since two hydrogen atoms and one oxygen have been removed to form water.

Complex carbohydrates such as starch, glycogen, and cellulose are longer chains formed in the same way as the disaccharides. For this reason, they are often called **polysaccharides**. Specific enzymes are required to form polysaccharides, each of which has a different arrangement of units. Since polysaccharides are not readily soluble in water, they provide a useful way for the cell to store energy. In general, plants form starches and cellulose, whereas animals

Figure 1.20.
The Structural and Simplified Structural Formulas for One Isomer of Glucose

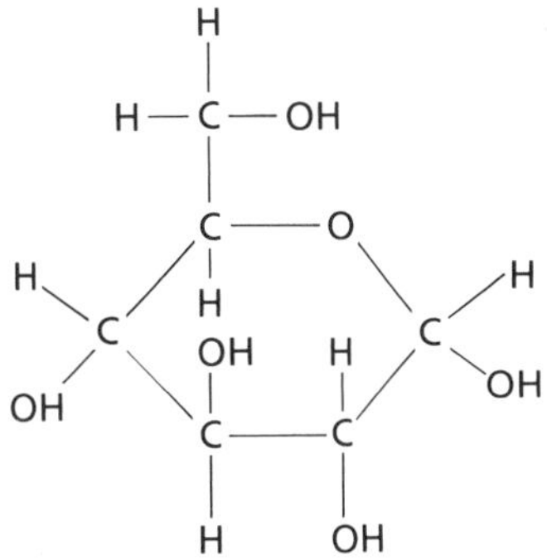

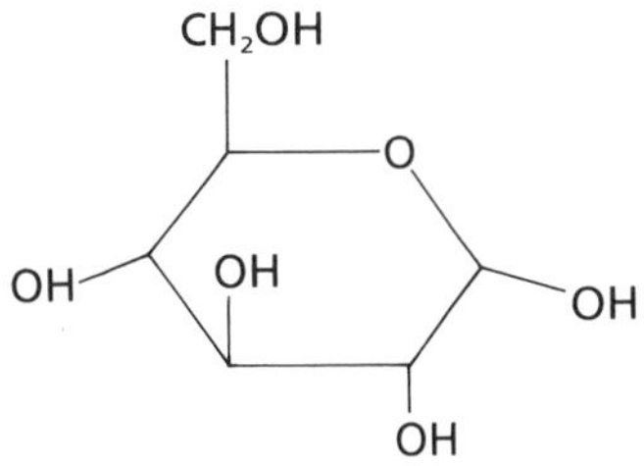

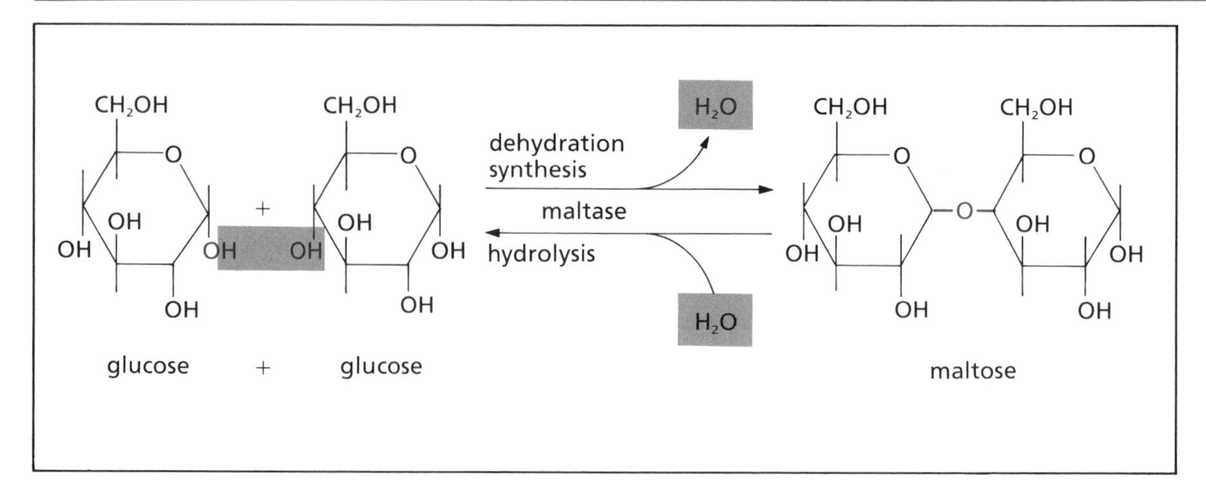

Figure 1.21.
The Synthesis and Digestion of a Disaccharide. When two glucose molecules are linked by dehydration synthesis, maltose is formed. Maltose units can then be linked together to form starch. Notice that this reaction is reversible; *i.e.*, maltose can be broken down, forming two glucose molecules.

synthesize glycogen. Humans can digest starches and glycogen, but not cellulose, since we lack the necessary enzymes. It is interesting to note that while people cannot digest cellulose (the fibre portion of the diet), the bacteria that are often found in the lower portion of the digestive system can.

Fats

Figure 1.22.
Acetic Acid: A Simple Organic Acid. The carboxyl group portion of the acetic acid (vinegar) molecule is shown in colour.

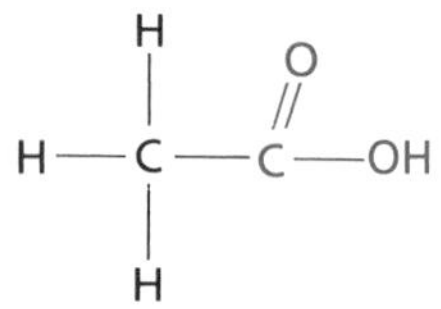

Fats are a different kind of polymer. Their basic building units are called **fatty acids**. Organic acids are characterized by the presence of a **carboxyl** group (COOH) in the molecule as, for example, in acetic acid (CH_3COOH), shown in Figure 1.22. Fatty acids are long chain molecules which, in the human body, usually contain an even number of carbon atoms.

The fatty acids in fats are not connected directly to each other. Instead, they are each linked, again by dehydration synthesis, to a molecule of **glycerol**. See Figure 1.23. When fats are broken down by hydrolysis, they form three molecules of fatty acids and one of glycerol from each fat molecule.

Proteins

Proteins, too, are polymers. They are made up of yet another type of unit — **amino acids**. Amino acids also contain the

glycerol + 3-stearic acid → tristearin + $3\,H_2O$

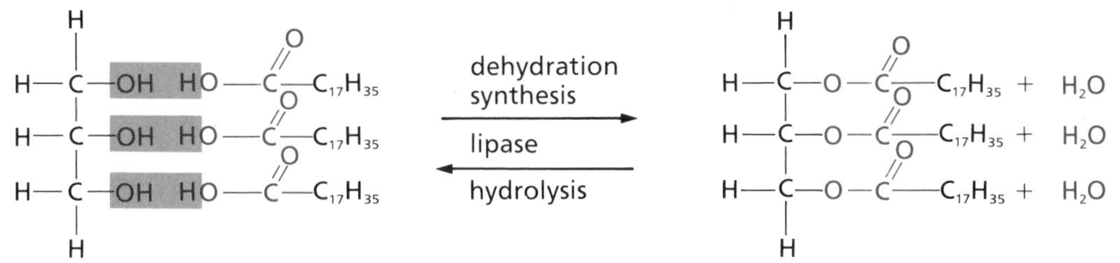

dehydration synthesis / lipase / hydrolysis

Figure 1.23.
The Synthesis and Digestion of a Fat. Three fatty acids are linked, by dehydration synthesis, to glycerol to form a fat. The three fatty acids need not be the same. Note that, like the reaction in Figure 1.21, this reaction is reversible.

carboxyl group typical of organic acids. In addition, however, they have another important group attached to them — the **amino group** (NH_2), which gives them their name. the structure of a simple amino acid is shown in Figure 1.24. There are twenty different amino acids which can be used to construct proteins. (They are listed in Chapter 4, Table 4.1.)

Amino acids, too, are linked together by dehydration synthesis. However, the linkage occurs not between two carbon atoms, but between the carbon of the carboxyl group of one amino acid and the nitrogen of the amino group of the other. This C–N linkage, shown in Figure 1.25 (on the next page), is called a **peptide bond**. Therefore, when two amino acids link together they are called a **dipeptide**; proteins may be called **polypeptides**. The enzymes which form and break these bonds are called **peptidases**.

Even the simplest proteins contain hundreds of amino acids, and unravelling their structure is the topic of much current research. The structure of insulin, a relatively small protein, is shown in Figure 4.2. In Chapter 4 you will learn how a cell can construct the immense variety of proteins it requires.

In spite of the amazing number of different chemicals found in cells, there are organizational patterns that enable cells to function very efficiently. As you will see in the next chapter, cells may use any of these substances to supply energy if necessary — yet only one common energy-releasing system is required.

Figure 1.24.
An Amino Acid: Glycine. The amino group is shown here in red. One of the hydrogen atoms attached to the central carbon atom can be replaced with various carbon chains to form other amino acids.

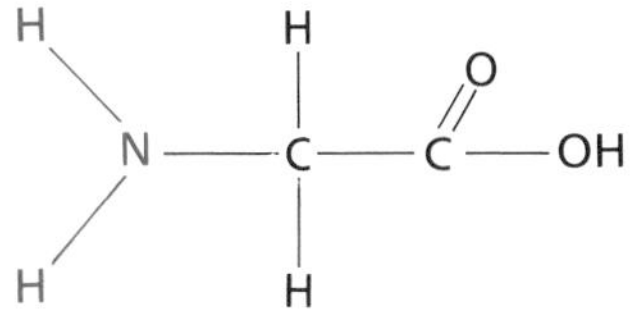

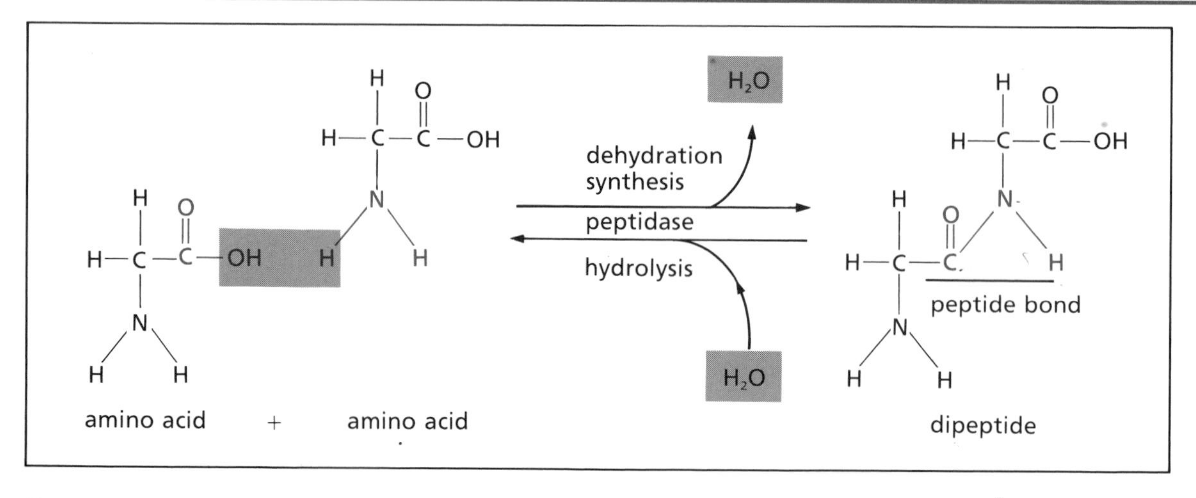

Figure 1.25.
The Synthesis and Digestion of a Dipeptide. Hundreds of amino acids can be linked in this way to form a protein.

QUESTIONS FOR REVIEW

SOME WORDS TO KNOW

Match each statement given in the left-hand column with a term shown in the right-hand column. DO NOT WRITE IN THIS BOOK.

1. The smallest unit into which an element can be divided and still retain its characteristics.
2. A substance composed of two or more elements in a definite proportion.
3. Atoms sharing electrons form a ______ compound.
4. A ______ speeds up a reaction.
5. A substance that contains just one kind of atom.
6. A positively charged particle in the atom.
7. The smallest unit into which a covalent compound can be divided.
8. After gaining or losing an electron, an atom becomes an ______ .
9. A bond in which electrons are shared unequally.
10. An ion present in the water solutions of all bases.
11. A substance that changes colour in an acid or base.
12. A ______ stabilizes the pH of solutions.
13. A diagram showing the arrangement of atoms and bonds in a molecule.

A. ion
B. hydrocarbon
C. catalyst
D. bonding site
E. polar
F. structural formula
G. lipids
H. protein
I. proton
J. covalent
K. element
L. indicator
M. molecule

14. A location on an atom where other atoms may be attached.	N. carbohydrate
15. A compound containing only carbon and hydrogen.	O. hydroxide
16. A carbon-based molecule in which the number of hydrogen atoms is twice the number of oxygen atoms.	P. compound
17. The general name for fats and oils.	Q. atom
18. A long chain organic molecule containing nitrogen.	R. buffer

SOME FACTS TO KNOW

1. Which four elements are most abundant in living things?
2. Why do living things need energy?
3. What determines the chemical behaviour of atoms?
4. Why are atoms electrically neutral?
5. Draw electron energy level diagrams for (a) nitrogen and (b) oxygen. Describe the outer energy level of each atom in terms of pairs of electrons and single electrons.
6. Write structural formulas for the compounds shown in Figure 1.5.
7. What characteristic of carbon atoms makes them suitable for forming chains and networks?
8. The fuel in some camp stoves is butane, C_4H_{10}. Draw the structural formula for butane.
9. State one important difference between the properties of ionic substances and covalent substances.
10. What is an ion?
11. Why do some substances dissolve in water?
12. Sulphuric acid is a strong acid whose formula is H_2SO_4. What happens to a molecule of H_2SO_4 when it reacts with a molecule of water? What is the name of the ion that is present in the water solutions of all acids?
13. By what factor would the hydronium ion concentration have to change to change the pH of a solution from 11 to 8?
14. What happens during neutralization?
15. What is the function of a buffer? List two examples of body systems which are very closely regulated by buffers.
16. When you eat a candy, the sugar in it is broken down first into glucose, then into carbon dioxide and water: $C_6H_{12}O_6 + 6O_2 \longrightarrow 6CO_2 + 6H_2O$. Estimate how much energy is released by the changes in chemical bonding in this reaction.
17. Removing a phosphate group from an ATP molecule results in release of energy for use in the cell. How can this be so when we know that all bond breaking requires an input of energy?
18. What are enzymes?
19. What factors can effect enzyme action?
20. How are polymers formed?

QUESTIONS FOR RESEARCH

1. The shape of a molecule is determined by forces that are predictable between the atoms composing it. Because these forces are predictable, it has been possible to develop computer programs to predict the shape and properties of complex molecules before they are synthesized. Investigate the applications of this technique in the chemical and pharmaceutical industries.

2. Water-softening devices operate by exchanging mineral ions, thus removing undesirable ones. Your kidney functions in a similar manner. Investigate how the required level of each ion is controlled.

3. The technique of blood dialysis, often called the "artificial kidney", filters the blood to restore a balance of ions that is as close to normal as possible. It is not as effective, however, as a healthy kidney. Investigate the causes of kidney deterioration. What are the limits of dialysis? What are the prospects for improving this technique? When is it advisable that a transplant be performed? How long does the average patient wait for a suitable organ?

4. Muscle contraction is now understood to be a chemical reaction initiated by changes in the distribution of sodium, potassium and calcium ions. Investigate the detailed chemistry of this process.

5. A great many household substances are dangerous chemicals. Parents and babysitters must be careful to keep these substances out of reach of children. Yet accidents do happen. Most major centres have now established poison information centres to provide advice and treatment in such emergencies. Where is your nearest poison information centre? How can you contact it? What services are provided? What are the most common causes of such accidents? How could they be prevented?

6. A special form of milk is now being introduced for those who are lactose intolerant. How does it differ from ordinary milk? Why is it more expensive? If you were in charge of marketing such a product, what ideas would you include in your advertising campaign? How successful do you think this product will be?

7. Compare a number of nonprescription vitamin and mineral supplements. What do they contain? Why do people need these substances? Evaluate these products in terms of their ability to supply the recommended daily allowance of these substances. How necessary are these products?

8. In many "diet" foods, synthetic chemical sweeteners have been replaced by sugars that are sweeter than sucrose (table sugar) but supply less energy. Investigate these substances. What is their source? How sweet are they? What is their chemical structure? How much energy do they release when metabolized? Are they safer than the substances they replace?

9. Use pH paper to determine the pH of household substances such as vinegar, ammonia, antacid tablets, juices, sour milk, drain cleaner, and others. The solids should be mixed with a little water before testing.

10. Design an apparatus that will allow a small flame to heat a quantity of water with minimum loss of heat to the surroundings. Compare the heat produced by a gram of burning candle wax to that produced by a gram of oil burning in a decorative lamp with a floating wick. Candle wax is a saturated hydrocarbon. Oil is an unsaturated hydrocarbon. Which do you think will produce more heat?

wilson's
ORGANIC
ROTENONE
INSECT DUST

DOMESTIC
500 g
REG. NO. 13693 P.C.P. ACT
GUARANTEE: Rotenone 1%
low toxicity organic insecticide for use on vegetables up to the day before harvest
READ THE LABEL BEFORE USING

H
DOMESTIC
NATURAL
INSECTICIDE
NATUREL
DOMESTIQUE

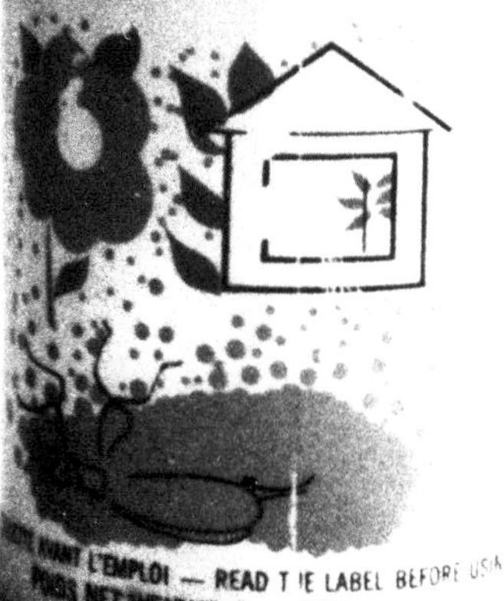

Chapter 2

Energy for Life

- Releasing Energy from Food
- Cellular Respiration

Chapter 2

Energy for Life

Every organism requires energy simply to keep its cells alive. Activity requires still further energy. Even as you read this, you are using almost 400 kJ/h of energy. If you stand up, you require more energy than when you were sitting; walking almost doubles your energy consumption. The various activities of work and recreation such as those shown in Figure 2.1 require even larger amounts of energy.

Figure 2.1.
The Energy You Use (Approximate Values)

Driving a Car	700 kJ/h
Walking normally	1000 kJ/h
Bicycling	2000 kJ/h
Downhill skiing, moderate speed	2500 kJ/h
Climbing stairs	3000 kJ/h

How does the cell obtain all this energy? Each cell has a supply of energy units in the form of molecules of adenosine triphosphate (ATP). The ATP molecule contains three phosphate groups. The third of these groups is attached by a bond which the cell can break as required to make use of the tremendous amount of energy it contains. When ATP releases this energy it becomes adenosine diphosphate (ADP), which has one less phosphate group. This process is illustrated in Figure 2.2. The second phosphate group is also attached by such a bond but this bond usually remains intact.

It is difficult to measure energy release in a cell, but approximately 30.66 kJ of energy are released by each mole of ATP that is converted to ADP. Under conditions of vigorous exercise, the cells involved must use this much energy each minute! Obviously, enormous numbers of ATP molecules are required by the body in the course of a day.

What happens when a cell runs out of ATP? Just as a car must be refueled periodically with organic molecules such

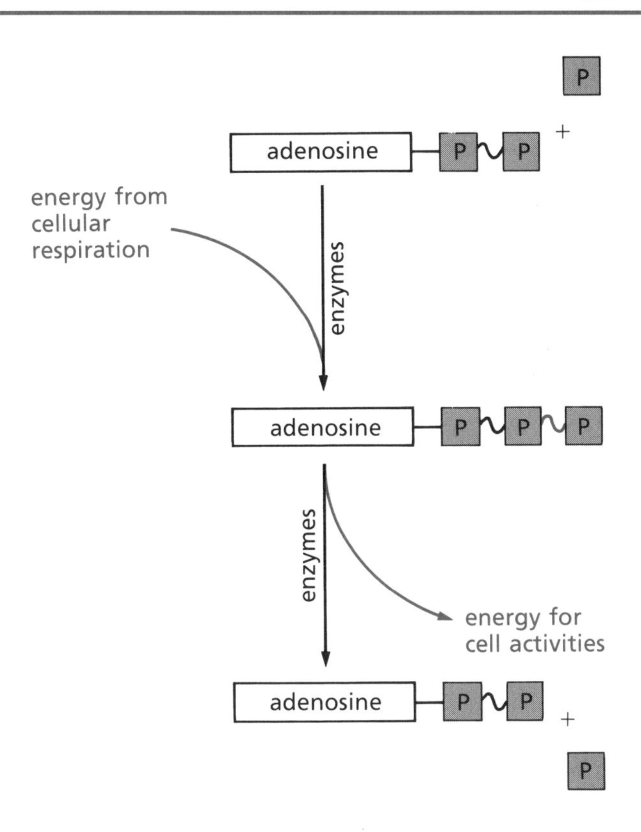

Figure 2.2.
ATP acts as a battery, storing energy from cellular respiration until it is needed by the cell for its activities. The wavy line indicates a high energy bond.

as those in gasoline or propane, the cell requires organic molecules as fuel. Glucose molecules are the major source of energy, but, if necessary, glycogen, glycerol, or amino acids may be used by some cells. The cell uses the energy contained by these molecules to manufacture more ATP through the process of **cellular respiration**. The cell's supply of molecules will eventually be consumed as well. How does it obtain more?

Our food that supplies the energy for life. If we trace that energy still further back, we discover that it originally came from the sun and was captured by a plant through the process of photosynthesis. Figure 2.3 (on the next page) summarizes these relationships. Green plants use the energy absorbed from sunlight to synthesize large molecules from simple inorganic molecules (mainly carbon dioxide and water). This process results in the formation of carbohydrates, fats, and proteins, which animals then consume as food.

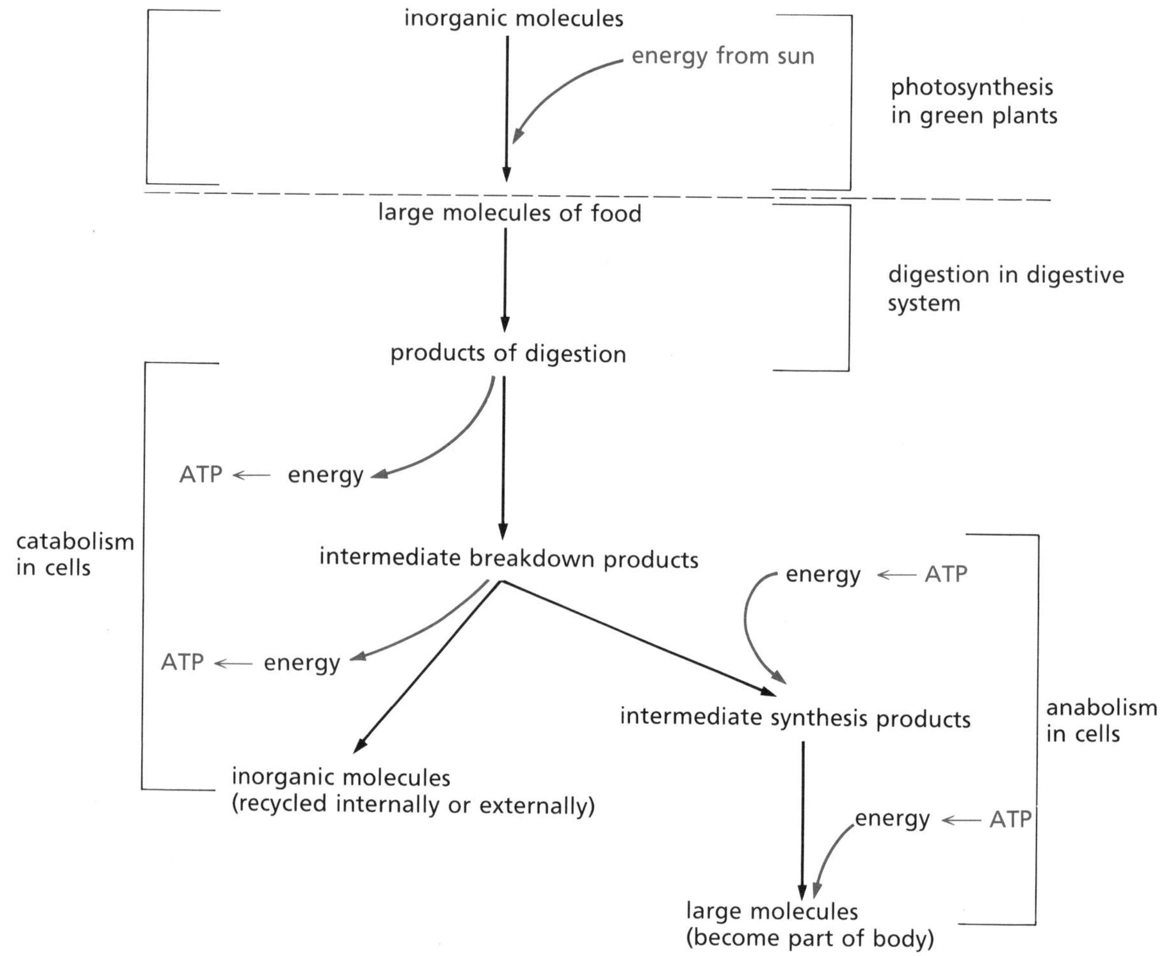

Figure 2.3.
Metabolic Pathways. Energy that came originally from the sun is consumed as food, then used to supply energy for cell activities.

Releasing Energy from Food

The energy which the food molecules contain is needed by the cells, but these molecules are too large to pass through cell membranes. The process of digestion (shown in Figure 2.4) is necessary to break down the food particles, first physically, then chemically, to a molecular size that can pass into the bloodstream for distribution. The carbohydrates are broken down into simple sugars called hexoses (primarily glucose); the fats are broken down into fatty acids and glycerol; the proteins are broken down into amino acids.

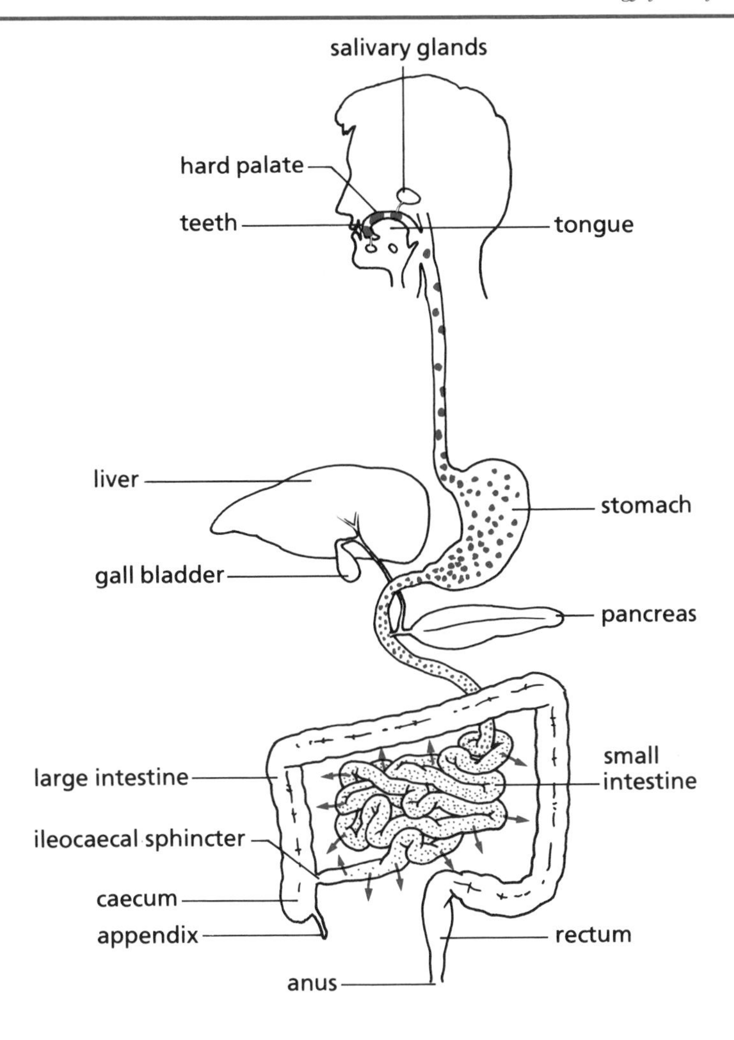

Figure 2.4.
The Role of the Digestive System. Large chunks of food enter the mouth. They are made smaller by the action of chewing. Physical breakdown is continued by the muscular action of the stomach. Chemical breakdown is started by the enzymes in the saliva and continues in the stomach and small intestine. The resulting molecules are absorbed through the walls of the small intestine.

These smaller molecules are then absorbed by the cells for immediate use, or stored until needed. They may be used as building blocks in the synthesis of larger molecules (**anabolism**), or may be broken down into inorganic molecules (**catabolism**) to release the energy they contain. Thus, the food you eat may be used to make new carbohydrates, fats, and proteins as required by each cell, or it may supply energy for such essential cellular activities as the active transport of substances through membranes, the transmission of nerve impulses, the synthesis of other molecules, the contraction of muscle cells, or movement.

Every cell must meet its own energy requirements by releasing energy from glucose or other molecules. Glucose is the most readily available material — its common name, "blood sugar", derives from the fact that it is normally present in the bloodstream. Red blood cells and brain cells utilize only glucose; they do not possess the enzymes necessary to metabolize other substances. Since glucose is a direct breakdown product of carbohydrate digestion, carbohydrates are the major energy source in a normal diet. Only if sufficient carbohydrate is unavailable will the liver act to convert other materials into glucose. Fats provide an alternative source of energy, since the products of their digestion may be used by many cells in cellular respiration.

Cellular Respiration

The complete combustion of a single gram of glucose releases 16 kJ of energy. The uncontrolled release of this much energy could destroy a cell! Therefore, the breakdown of the glucose molecule must be regulated by enzymes and accomplished step by step. As Figure 2.5 shows, about 40% of the energy obtained from a glucose molecule is used for converting ADP to ATP. While this appears to be rather inefficient, the process of cellular respiration is actually more efficient than the average car engine.

Figure 2.5.
The Efficiency of Cellular Respiration. Forty percent of the chemical energy in the glucose molecule is converted to useful energy in the form of ATP. The remaining 60% is released as other forms of energy.

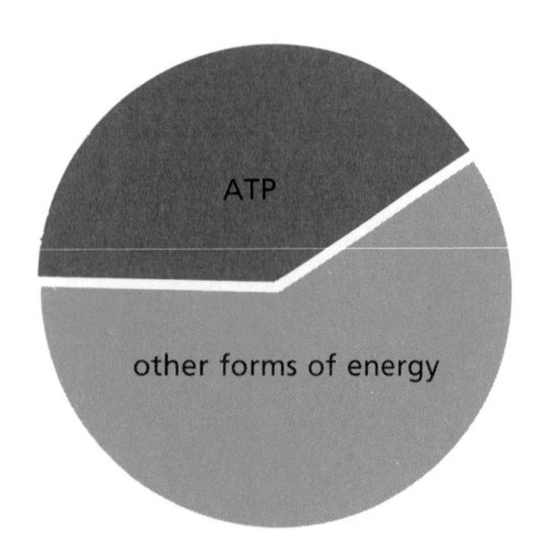

Cellular respiration involves a series of chemical reactions, each controlled by a specific enzyme. They are usually considered in groups, as you can see in Figure 2.6.

The first stage in the release of energy from glucose is called **glycolysis**. During this stage, glucose is broken down into pyruvic acid. This breakdown takes place in the cytoplasm of the cell, and is said to be **anaerobic**, since it does not require oxygen. Because small amounts of ATP are formed during this stage, cells can survive short periods of oxygen deficiency.

The remaining stages occur only if oxygen is available; thus, they are said to be **aerobic**. The further breakdown of the pyruvic acid molecule is accomplished through the reactions of **Krebs' Cycle**, named after Hans Krebs, who received a Nobel Prize for his discovery. (This cycle is sometimes also called the citric acid cycle or the tricarboxylic acid cycle.) The reactions of Krebs' Cycle are controlled by enzymes arranged along the membranes of the cell's mitochondria.

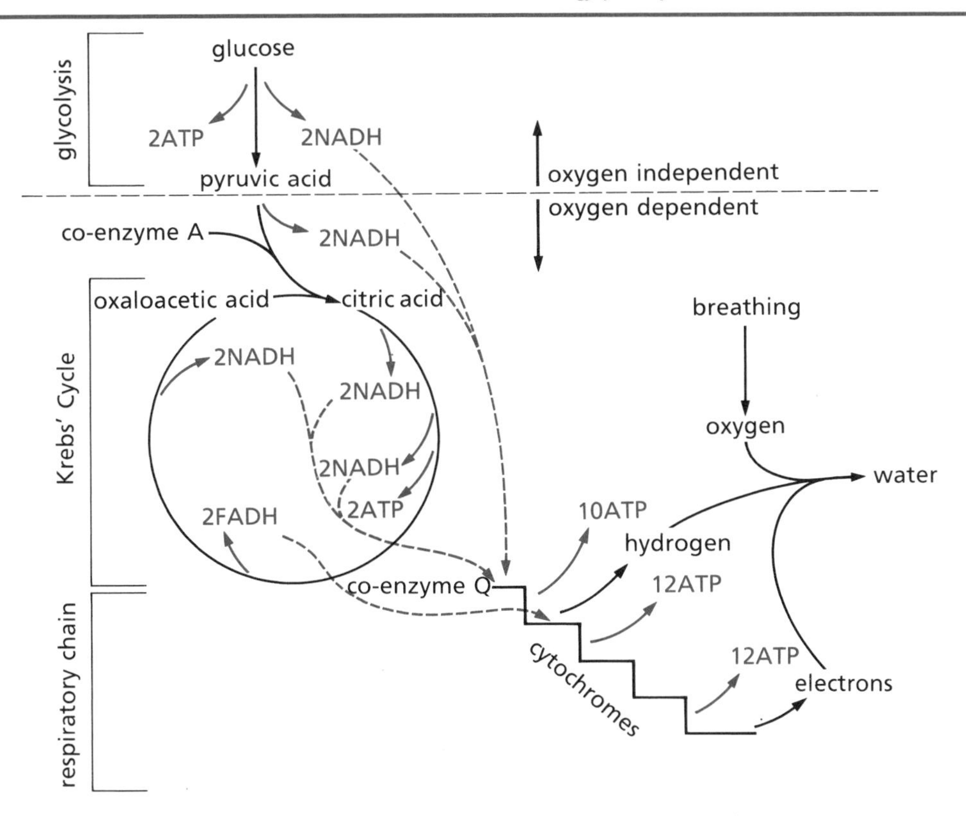

Figure 2.6.
The energy releasing processes of the cell occur in three main groups: glycolysis, Krebs' Cycle, and the respiratory chain.

Although at some stages in glycolysis and Krebs' Cycle the quantities of energy released permit the direct synthesis of ATP, at other stages the energy is carried by a "hot" electron which has been removed. These electrons are transferred to the **respiratory chain**, another series of enzymes in the mitochondria. Here, the energy of each electron is gradually released and used to manufacture more ATP. Each of these processes is discussed in more detail below.

Glycolysis

This first stage in the breakdown of the glucose molecule is outlined in Figure 2.7 (on the next page). Just as you must light a fire in order to release the heat energy contained by the fuel, the cell must supply the activation energy needed to initiate the breakdown of the glucose. The energy is supplied by attaching the phosphate group from a molecule of ATP to the glucose, turning it into **glucose phosphate**. The cell then rearranges this new molecule into a different isomer called **fructose phosphate**.

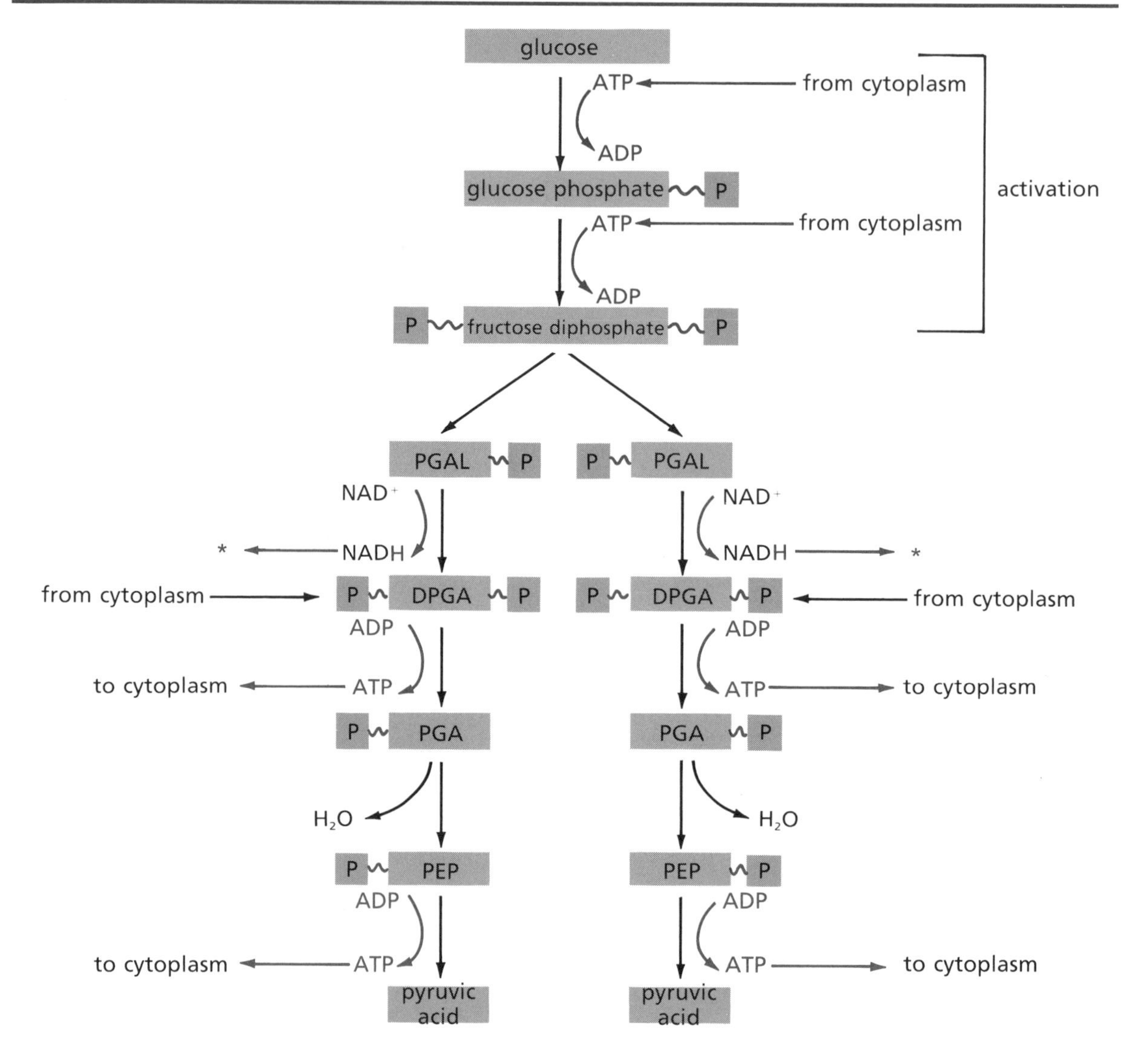

Figure 2.7.
The Stages of Glycolysis. During this process, a six-carbon molecule of glucose is broken down into two three-carbon fragments. The phosphate groups, shown in red, transfer energy.

Still more energy is required to start the reaction, so a second phosphate group is transferred from ATP to form **fructose diphosphate**. Now that it has been activated sufficiently, this six-carbon molecule can be split by the cell into two three-carbon fragments, as shown in Figure 2.7. These are **phosphoglyceraldehyde** (usually written as PGAL), and an isomer of it which can also be used in glycolysis.

Now the process of stripping the energy from these molecules can begin. A high energy hydrogen ion is removed from each PGAL. Simultaneously, a phosphate group is added to each PGAL to form **diphosphoglyceric acid** (DPGA). This phosphate comes from inorganic phosphate in the cytoplasm, not from ATP.

The hydrogen ions that were removed are accepted by the co-enzyme **nicotinamide adenine dinucleotide** (NAD^+), which then transfers them to the respiratory chain for further energy release. This essential co-enzyme is synthesized from vitamin B_3 (niacin or nicotinic acid); therefore, you must be sure that you eat foods containing this vitamin. A list of the foods that are excellent sources of B vitamins can be found in Appendix 1.

The cell then uses the phosphate groups to transfer energy to ADP, thus forming two molecules of ATP. At this point, the ATP needed to start the process has been replaced, and two molecules of **phosphoglyceric acid** (PGA) remain.

The cell now converts the PGA to another isomer and removes a molecule of water from each one. This results in the formation of **phosphoenol pyruvic acid** (PEP). In PEP molecules, the phosphate group has much higher energy than it did in PGA. The cell now removes the phosphate group from each PEP molecule, again using it to convert ADP to ATP. This leaves two molecules of **pyruvic acid** and completes the process of glycolysis.

Glycolysis thus results in a net yield of two molecules of ATP, which the cell can use to supply energy for other processes. Since each of these reactions is catalyzed by a specific enzyme, the cell retains tight control over the step-by-step release of energy.

Breakdown of Pyruvic Acid

The type of organism involved and the availability of oxygen determine what happens to the pyruvic acid. In most plants and many microorganisms, if oxygen is deficient, the process of **fermentation** occurs, resulting in the formation of alcohol. During fermentation, a molecule of carbon dioxide is removed from the pyruvic acid, and the pyruvic acid is converted into ethanol using the hydrogen ion which had been transferred to NAD^+.

Animal cells and other microorganisms carry out a similar process, known as **lactic acid fermentation**, if the oxygen supply is low. It, too, uses the hydrogen ion that had been transferred to NAD$^+$, but converts the pyruvic acid into lactic acid. Although both of these processes use energy, their value to the cell lies in the fact that they free the NAD$^+$ to accept more hydrogen ions, thus enabling glycolysis to continue to supply at least some energy to the cell. If this did not occur, hydrogen ions would accumulate, making the cell so acidic that it might die.

Muscle cells enter lactic acid fermentation during vigorous or prolonged activity when the circulation cannot deliver sufficient oxygen to meet the demand. The gradual accumulation of lactic acid poisons the cells and interferes with their activity, resulting in the effect known as "cramps". Some athletes are experimenting with doses of hydrogen carbonate ions before activity to delay the onset of the symptoms of lactic acid accumulation. When the muscle is allowed to rest, the lactic acid is carried by the bloodstream to the liver, where it is converted to glycogen. This conversion can occur only when sufficient oxygen is available; thus, this process is sometimes called the **oxygen debt mechanism**. Athletes who breathe heavily after exertion, like the ones in Figure 2.8, are repaying the oxygen debt.

If sufficient oxygen is present, however, these processes do not occur, and the breakdown of the pyruvic acid continues in the mitochondria. When it enters a mitochondrion, the pyruvic acid molecule is immediately stripped of a hydrogen ion. Simultaneously, a molecule of carbon dioxide is removed. The hydrogen ion is accepted by NAD$^+$ for transfer to the respiratory chain. The remaining two-carbon fragment is picked up by **co-enzyme A** to form a temporary complex which enters Krebs' Cycle. This process requires an adequate supply of vitamin B_1 (thiamine).

Figure 2.8.
Repaying an Oxygen Debt. The body requires additional oxygen to convert the lactic acid accumulated in the muscles into more useful glycogen. The greater the level of exertion and the longer its duration, the longer the period of heavy breathing which follows.

Krebs' Cycle

This stage of cellular respiration completes the breakdown of the two-carbon fragment remaining from the glucose. It is accomplished by attaching the fragment to **oxaloacetic acid**, a four-carbon molecule. This process requires the addition of a molecule of water and releases the co-enzyme A for further activity. The resulting six-carbon molecule is **citric acid**, which is why this cycle is often given that name. Figure 2.9 illustrates how the oxaloacetic acid is regenerated as the energy still remaining in the fragment is released.

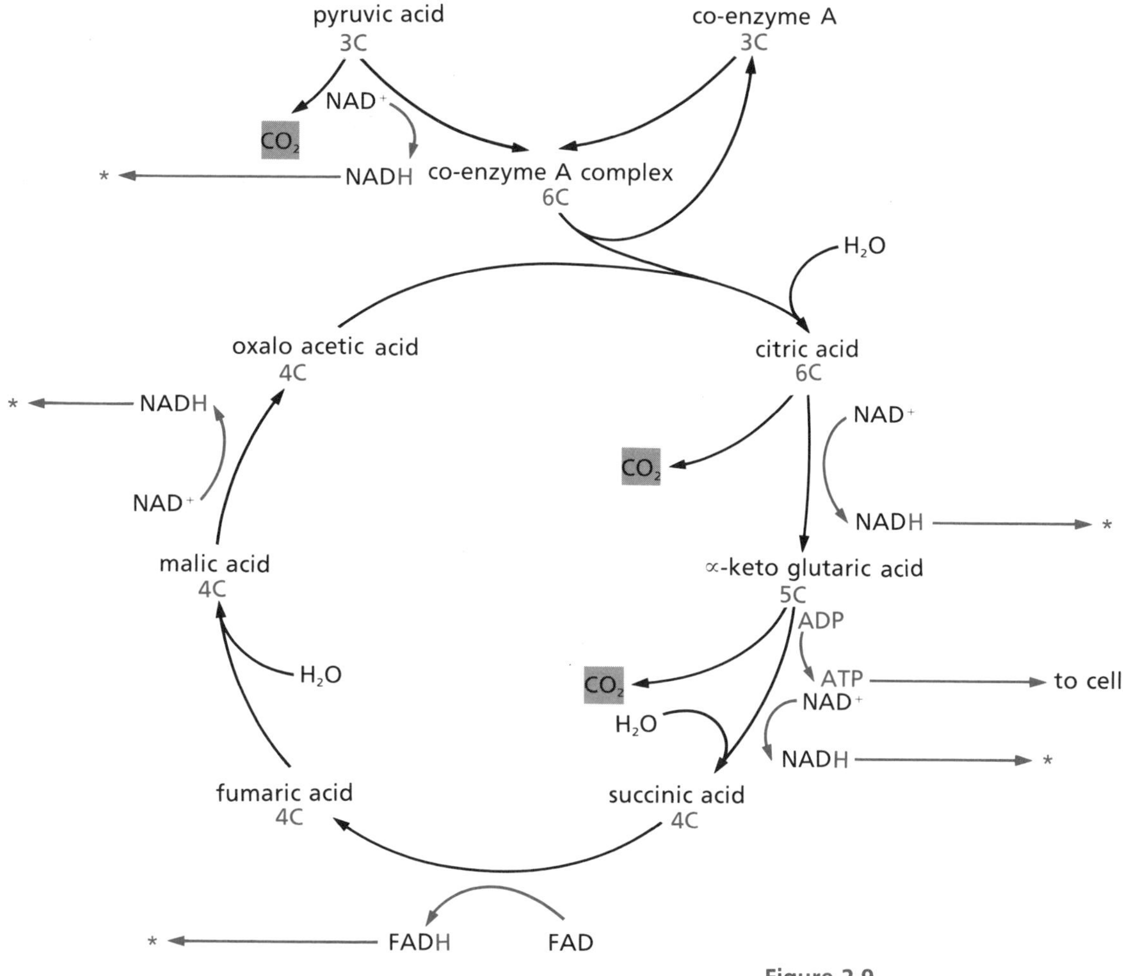

Figure 2.9.
The Stages of Krebs' Cycle. Most of the energy resulting from this cycle is not released directly. The energy is transferred by co-enzymes to the respiratory chain, where it is released, as shown in Figure 2.6.

Two successive reactions each remove a hydrogen ion and a molecule of carbon dioxide from the citric acid. Then a molecule of water is added to the remaining fragment to form **succinic acid**. During these reactions, a molecule of ADP is converted to ATP and the two hydrogen ions are transferred to the respiratory chain by NAD^+.

Another hydrogen ion is then removed from the succinic acid. Since it does not carry quite as much energy as the first, it is transferred, not to NAD^+, but to another co-enzyme, **flavin adenine dinucleotide** (FAD). This co-enzyme is also formed from a B vitamin, vitamin B_2 (riboflavin).

The succinic acid has now become **fumaric acid**, which is then converted to **malic acid** by the addition of a molecule of water.

Now the remaining hydrogen ion is transferred to NAD$^+$, resulting in the regeneration of the oxaloacetic acid. Thus, the pyruvic acid has been split to form two molecules of carbon dioxide, with four hydrogen ions being transferred to the respiratory chain.

Since each molecule of glucose produced two molecules of pyruvic acid, in total, six molecules of ATP have been produced directly (a net gain of four). There are, however, ten molecules of NADH and two of FADH still to release their energy through the reactions of the respiratory chain. It is interesting to note the major role played by the B vitamins in these energy releasing processes. They are required for the formation of three essential co-enzymes; therefore, a deficiency of any one of these vitamins can interfere with the entire process. Fortunately, we can be sure that we consume enough of these vitamins if we eat a properly balanced diet, as outlined in Canada's Food Guide. The table in Appendix 2 identifies foods that are particularly rich in specific vitamins.

Figure 2.10.
The Stages of the Respiratory Chain. Here the high energy electrons are transferred to successively stronger electron acceptors. This system accepts electron carrying co-enzymes from various stages.

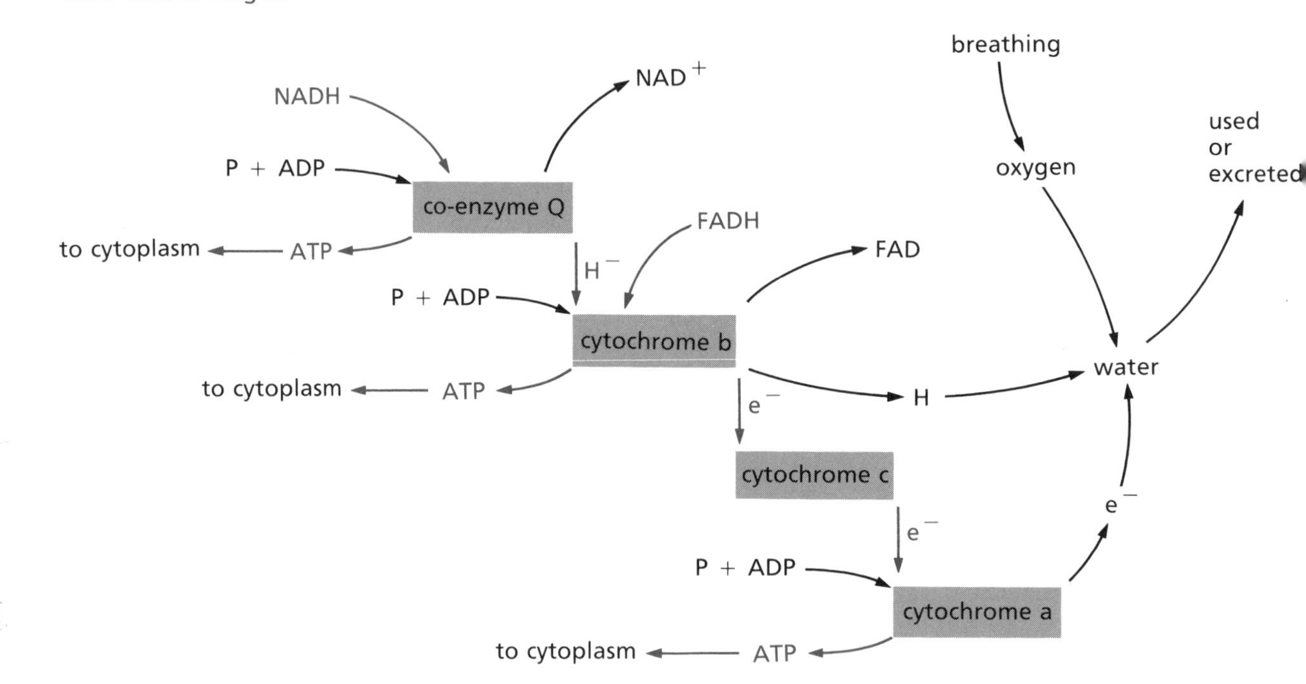

The Respiratory Chain

Most of the energy originally contained in the glucose still remains to be processed. It has been transferred temporarily to the co-enzymes NAD$^+$ and FAD by hydrogen ions. As Figure 2.10 demonstrates, this energy must be gradually released through a series of step-by-step reactions. These reactions are controlled by a series of enzymes and co-enzymes arranged along the membranes of the mitochondria. The structure of this organelle is shown in Figure 2.11.

As the hydrogen ions are passed to successively stronger electron acceptors, the energy they carry is used to synthesize ATP from ADP and inorganic phosphate. The hydrogen ions carried by NAD$^+$ are transferred first to **co-enzyme Q**. Then, as they pass from co-enzyme Q to **cytochrome b**, a molecule of ATP is synthesized using the energy that is released. The hydrogen ions carried by FAD are transferred directly to cytochrome b. Thus, each of these hydrogen ions produces only two molecules of ATP, not three, as it passes down the chain.

Only the electron is transferred from cytochrome b to **cytochrome c** and then to **cytochrome a** during the next reactions, which result in the formation of two more molecules of ATP. The electron, now stripped of all its extra energy, is returned from cytochrome a to the hydrogen atom released by cytochrome b. This hydrogen is then disposed of by being combined with oxygen to form water.

The major function of the oxygen we obtain from breathing is to remove the hydrogen remaining from cellular respiration. Scientists are now studying organisms that can carry out cellular respiration efficiently in the absence of oxygen. These organisms use sulphur as an alternative means of disposal, producing hydrogen sulphide instead of water as a waste product.

During the reactions of the respiratory chain, each molecule of NADH that is transferred to the system results in the formation of three molecules of ATP, for a total of thirty. Each molecule of FADH that passes down the chain produces two molecules of ATP, for a total of four. Thus, thirty-four molecules of ATP are produced in the respiratory chain, to be added to the four that were produced during glycolysis. The net yield is thirty-eight molecules of ATP from each molecule of glucose that enters glycolysis.

Figure 2.11.
The Structure of a Mitochondrion. The enzymes and co-enzymes of the respiratory chain are thought to be arranged along the membranes of the mitochondria in sequence.

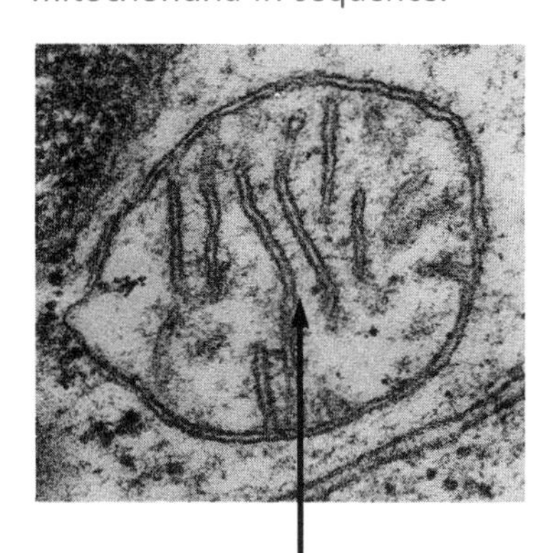

Other Pathways

The processes of cellular respiration do not occur in isolation from the other reactions of the cell. Materials are transferred between reactions according to the requirements of the cell at the time. Figure 2.12 shows some of the possibilities. If more energy is required than can be supplied directly by glucose, the liver can convert its stored glycogen to glucose phosphate, enabling it to enter the system. The body stores enough glycogen for about half a day of normal activity, but a marathon runner would consume this supply in about an hour and a half.

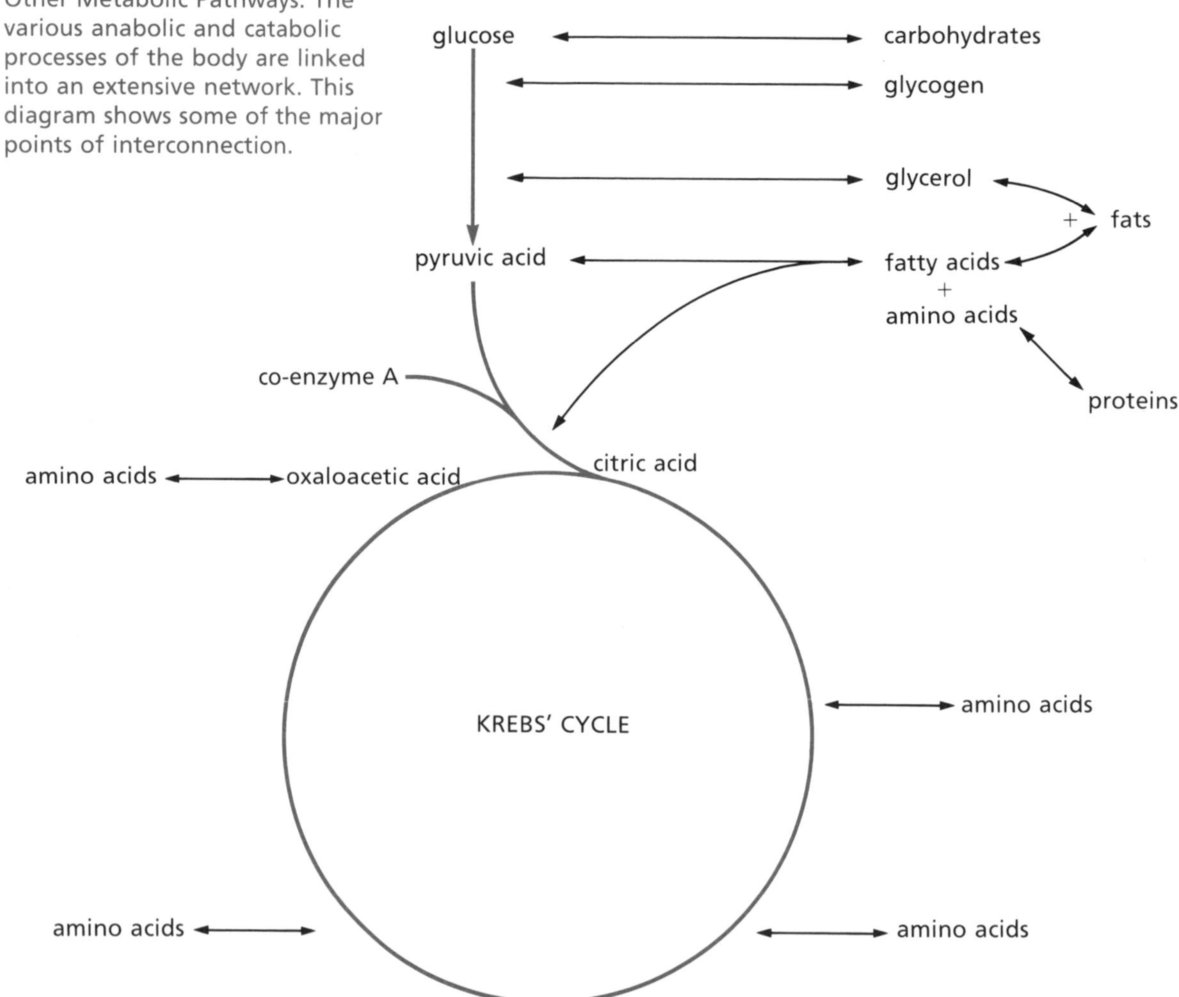

Figure 2.12.
Other Metabolic Pathways. The various anabolic and catabolic processes of the body are linked into an extensive network. This diagram shows some of the major points of interconnection.

The glycerol that results from the breakdown of fats enters glycolysis as PGAL. The resulting fatty acids can enter the system at various points, as can the amino acids resulting from protein breakdown. Excess amino acids are used in this way, but the cell will only break down its own protein as a source of energy under conditions of starvation, or deliberate extreme dieting, after all other sources of energy have been used. Protein is a poor source of energy, and the resulting waste products put a strain on the kidney.

Excess materials may also be removed from the cellular respiration system for temporary storage. Glucose may be converted to the more complex carbohydrate, glycogen. PGAL may be diverted to form fats in combination with fatty acids synthesized from pyruvic acid. The pyruvic acid molecules may also be used to manufacture various amino acids for protein synthesis. Thus the fate of a given molecule depends on which enzyme reaches it first!

Although each cell must meet its need for energy through its own chemical processes, it must also function in coordination with the other cells of the body. The level of energy release in each cell must be suited to the current needs of the entire organism. In turn, the body's transportation system must deliver and remove materials at a suitable rate if the cells are to function effectively. Thus, the processes we have been considering in this chapter must be controlled by another system in addition to the enzymes of each cell. This network of controls forms the subject of Chapter 3.

QUESTIONS FOR REVIEW

SOME WORDS TO KNOW

Match each description given in the left-hand column with a word shown in the right-hand column. DO NOT WRITE IN THIS BOOK.

1. The energy unit within the cell.
2. A major source of energy supplied to the cell.
3. The process that first breaks down food.
4. A process that forms larger molecules.
5. A substance that controls a biological reaction.
6. The first stage of energy release within a cell.
7. Means "without oxygen".
8. A process that breaks down pyruvic acid.
9. The location of the final stages of energy release within the cell.
10. A co-enzyme which accepts hydrogen ions.
11. An important electron acceptor.

A. Krebs' Cycle
B. enzyme
C. cytochrome
D. ATP
E. anaerobic
F. NAD^+
G. glycolysis
H. glucose
I. anabolism
J. mitochondrion
K. digestion

SOME FACTS TO KNOW

1. How does the cell store energy in ATP?
2. What molecules can the cell use to manufacture more ATP?
3. Why must food first be digested before it can be utilized?
4. What happens to the energy in glucose that is not used to make ATP?
5. Why is ATP required to initiate glycolysis?
6. What is accomplished during glycolysis?
7. What is the role of NAD^+?
8. Under what conditions does lactic acid fermentation occur?
9. What is accomplished during Krebs' Cycle?
10. What is the function of the respiratory chain?
11. What is the role of oxygen in cellular respiration?
12. Why are there alternative pathways associated with cellular respiration?

QUESTIONS FOR RESEARCH

1. Many athletes, immediately before or during competition, consume large quantities of glucose in either tablet or beverage form. Determine whether this practice is advantageous. What undesirable effect(s) might it have?

2. Patients who are unable to consume food must be fed intravenously. Find out what an intravenous solution would need to contain for a patient to regain health.

3. Some athletes now engage in the practice of "bicarbonate loading" before their event. Determine whether this practice is effective. In what type of event might it confer an advantage?

4. The typical North American diet is said to contain far more protein than necessary. Find out whether this statement is valid. What happens to the excess protein? In general, protein is more expensive than other nutrients. What could replace much of it in our diet?

5. Sports nutritionists now recommend a meal including pasta just prior to an event rather than the traditional steak and salad. Investigate the basis for this recommendation.

6. The description of the metabolic pathways presented in this chapter has been very much condensed. Obtain a modern biochemistry text and research some of these pathways in more detail.

7. The molecular changes involved in metabolism are fascinating. To help you to visualize what is taking place, construct models of the molecules involved. A group might work together to produce models of an entire process.

LANGSTAFF
GYMNASTIC CLUB

Chapter 3

The Body in Balance

- Feedback Systems
- Regulation of Metabolism
- Regulation of Body Temperature
- Regulation of Blood Glucose
- Regulation of Digestion
- Regulation of Breathing
- Regulation of Other Systems
- Upsetting the Balance

Chapter 3

The Body in Balance

Just as gymnasts must make constant adjustments to maintain balance as they move, so your body must make constant adjustments to keep its internal environment stable while its external environment changes. These processes occur automatically — you do not consciously "adjust your thermostat" when you move from the chill of the ski slopes to the warmth of a fireplace; you do not have to reset your metabolism if, after skipping breakfast, you have a honey doughnut for a mid-morning snack. When you do have breakfast, your digestive system copes equally well with grapefruit juice of pH 3 and eggs of pH 8 without your assistance.

What makes all these adjustments possible? How does your body keep its internal environment so finely balanced? These regulatory processes depend upon information received from a wide variety of sensors distributed throughout the body. They are coordinated by the action of nerve impulses, enzymes, and hormones. (A hormone is a chemical messenger which is released by one organ to travel through the bloodstream to act on a target organ.)

Feedback Systems

Each major body process is organized in such a way that it is self-regulating. The controlling system constantly receives **feedback** (information) as to whether body conditions are normal. The activity level of the process increases or decreases, as necessary, to maintain conditions within the range required for normal functioning. This type of process is called a **homeostatic mechanism** or a **feedback system**. Thus, homeostasis is a condition of active balance in which body processes fluctuate within a range considered normal.

For each feedback system to operate, three components are necessary: (a) a **monitor** which detects the change; (b) a **control centre** which selects an appropriate adjustment;

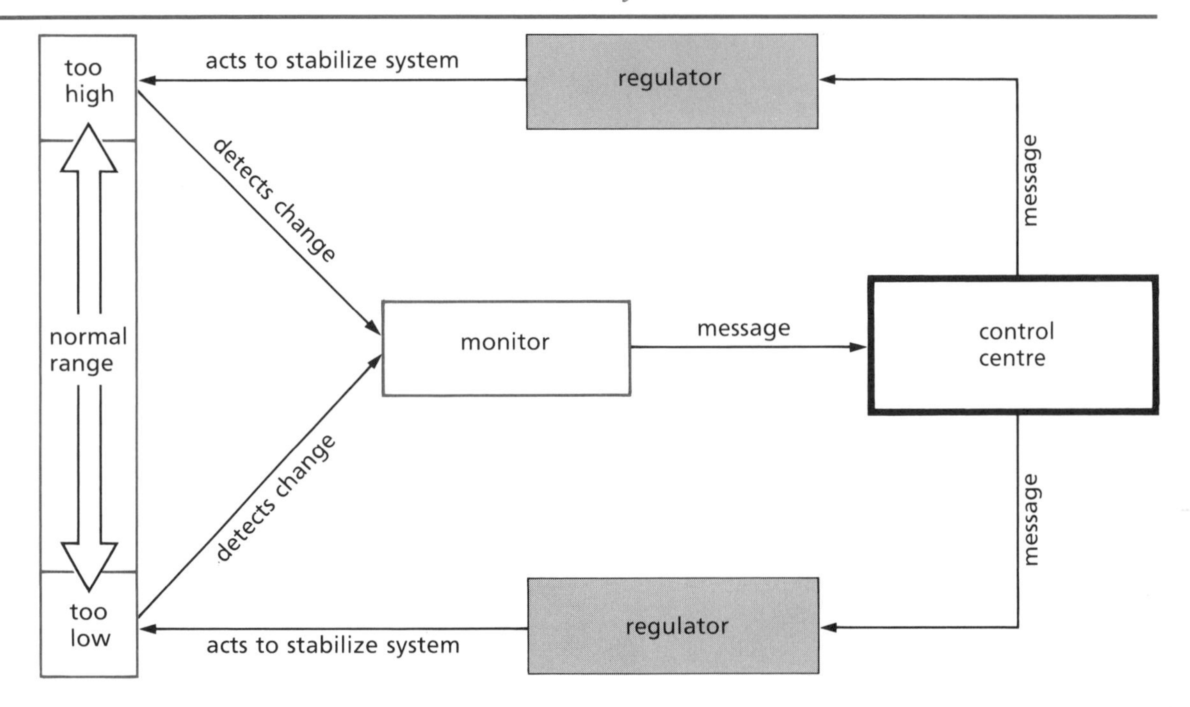

Figure 3.1.
The Components of a Feedback System. Each component is differently coded. This pattern will be retained in the illustrations of feedback systems that follow.

and (c) a **regulator** which carries it out. The messages transmitted between these three may be chemical or nervous in nature. A generalized version of such a system is shown in Figure 3.1.

The limits of the range of activity which may be considered "normal" vary from system to system. The monitor sends a message to the control centre only when the system starts to move beyond its normal limits and thus threatens homeostatic balance. The control centre then initiates a correcting process to return the system to balance. The net result is that the system fluctuates above and below an "ideal" level, but within the limits of normality. Some specific examples of these regulatory processes will illustrate the patterns of control which keep your body functioning normally under a wide variety of conditions.

Regulation of Metabolism

The **metabolic rate**, the rate at which cells carry out chemical activities, is under the control of a feedback system. Figure 3.2 (on the next page) illustrates the process. The **hypothalamus** portion of the brain, shown in Figure 3.3

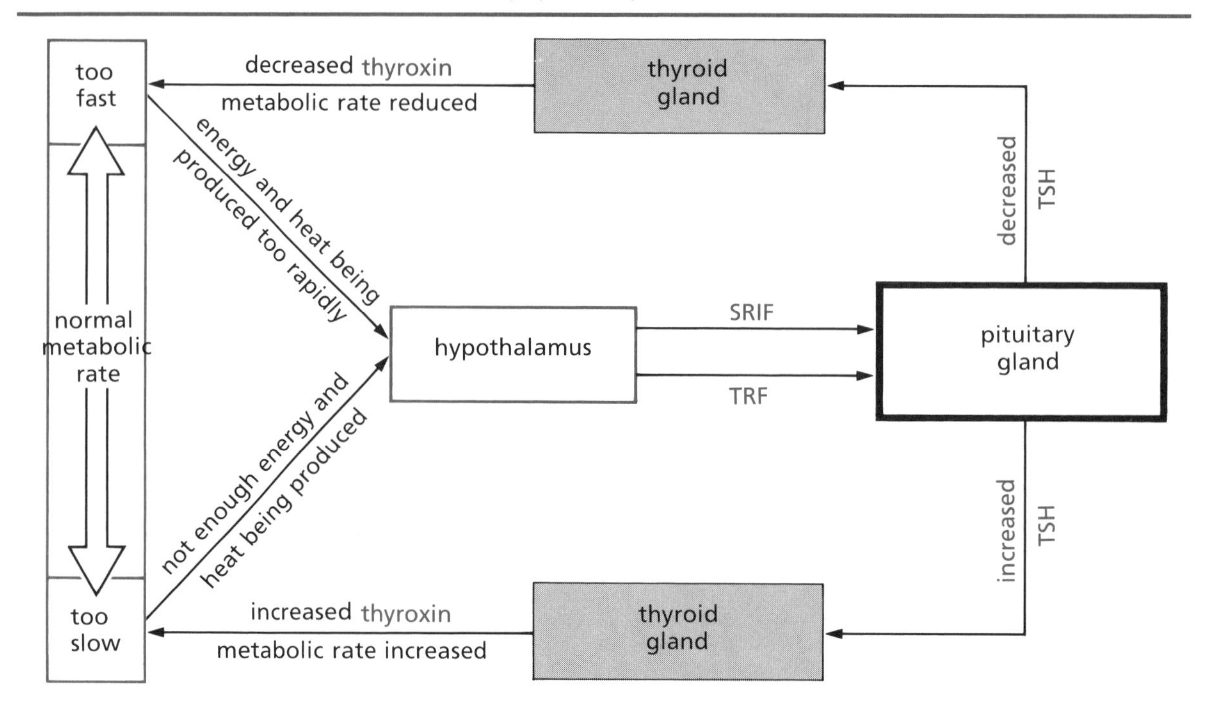

Figure 3.2.
The Feedback System Controlling Metabolic Rate. Note the messages sent by the hypothalamus to the pituitary gland to signal the switching on and off of thyroxin production.

SRIF = Somatotropin Release Inhibiting Factor
TSH = Thyroid Stimulating Hormone
TRF = Thyrotropin Releasing Factor

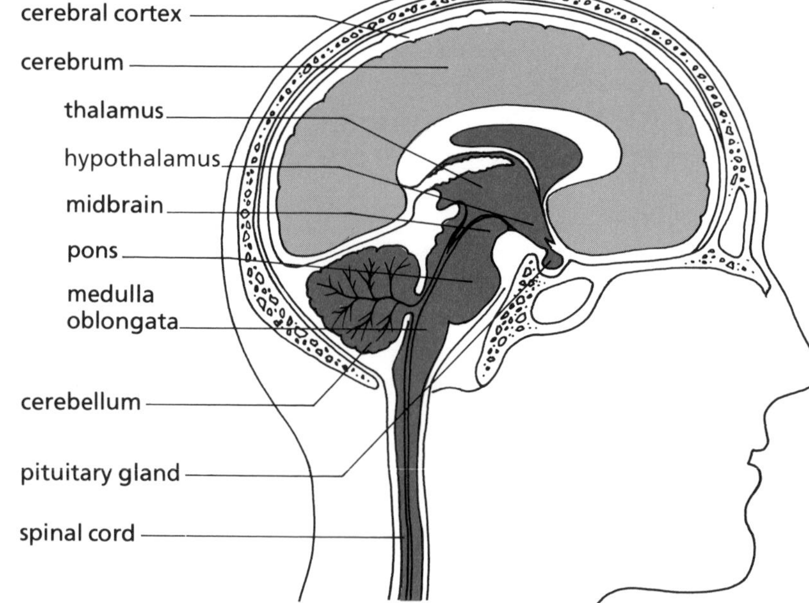

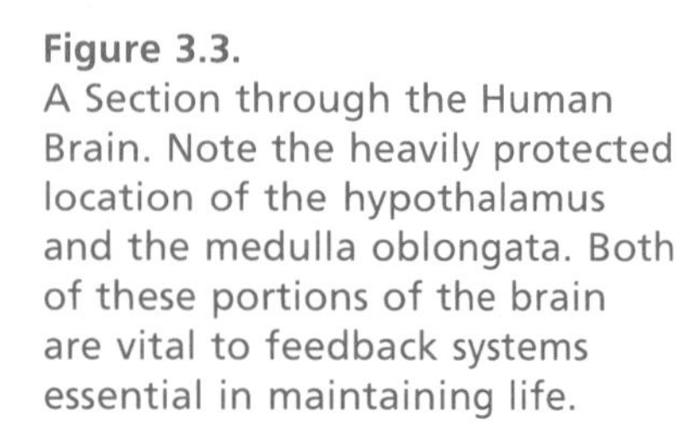
Figure 3.3.
A Section through the Human Brain. Note the heavily protected location of the hypothalamus and the medulla oblongata. Both of these portions of the brain are vital to feedback systems essential in maintaining life.

(on the next page), monitors the rate of production of heat and energy. If the metabolic rate is too high, the hormone **somatotropin release inhibiting factor** (SRIF) is sent to the pituitary gland in the brain, the control centre for this system. The pituitary gland, in turn, reduces its output of another hormone, **thyroid stimulating hormone** (TSH). The reduction in TSH causes the thyroid gland to reduce its output of the hormone **thyroxin**. The result is a reduction of the rate of metabolism in the cells.

If, on the other hand, the metabolic rate is too low, the hypothalamus sends the hormone **thyrotropin releasing factor** (**TRF**) to the pituitary gland. The result is an increased output of TSH from the pituitary gland. This increase then causes the thyroid gland to release more thyroxin, thus increasing the metabolic rate.

This feedback system is actually part of a much more complex system responsible for maintaining a stable internal temperature.

Regulation of Body Temperature

Cells can function normally only within a very narrow range of temperatures, so our internal temperature variations must be stabilized rapidly. Yet we can function within a wide range of environmental temperatures without succumbing to hypothermia or heat exhaustion. How is this possible?

Consider the system outlined in Figure 3.4 (on the next page). Two sets of monitors exist. One set, in the skin, monitors surface temperature and feeds that information to the hypothalamus. The rest of the monitoring system lies within the hypothalamus itself; it monitors the temperature of the blood reaching the brain. Since immediate correction is essential, the hypothalamus (the monitor) does much more than merely "adjust the thermostat" by means of the thyroid gland (the regulator).

Response to Low Temperatures

The following analogy may help you understand the action of the hypothalamus. If it is chilly when you return home, it can take a long time for the furnace to produce its effect. While you wait, you might light a fire or turn on a heater to supply immediate heat. The hypothalamus initiates a similar action if the blood temperature is too low.

Muscular activity results in considerable output of heat

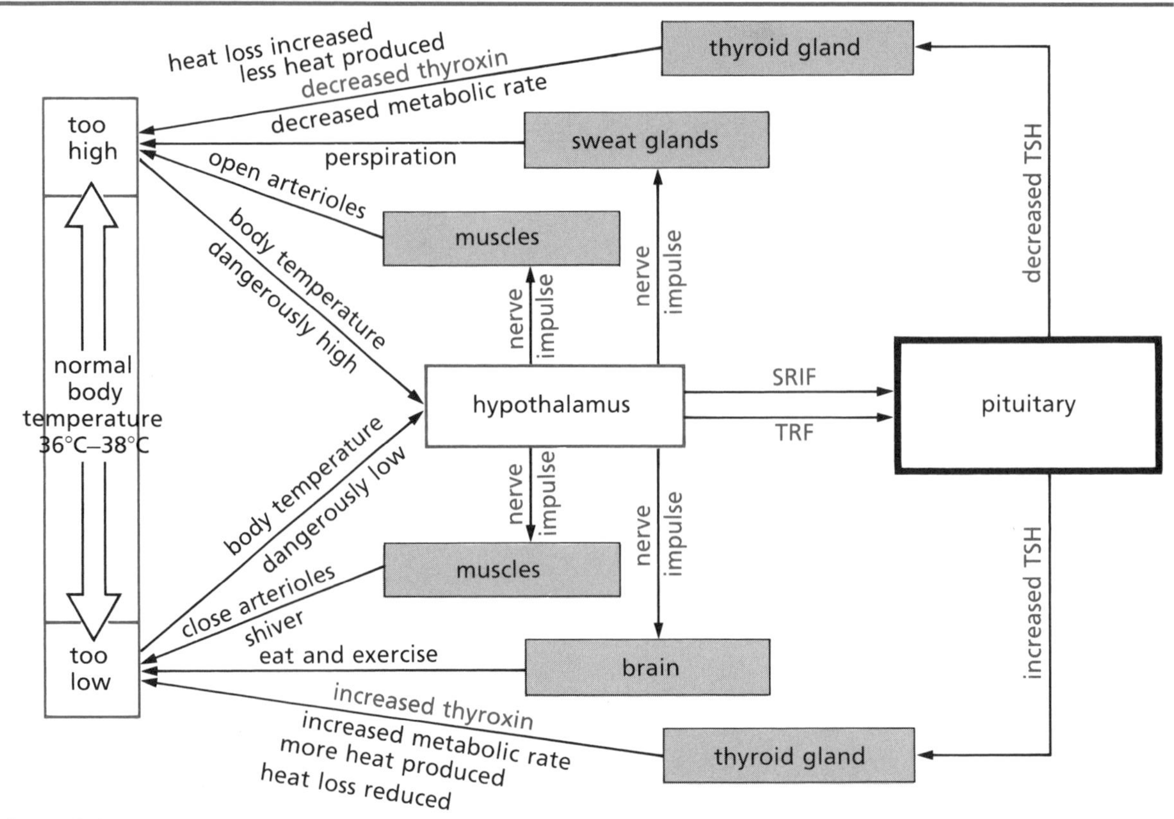

Figure 3.4.
The Systems Controlling Body Temperature. Since a constant body temperature is essential to normal functioning, adjustment is made at two levels. Both heat production and heat loss from the body surface are controlled.

energy. Thus, the hypothalamus sends messages to other portions of the brain, telling the body to become more active. If this is ineffective, it will send messages directly to the muscles, causing repeated contractions — the effect we know as shivering. This intense muscular activity produces the heat needed to raise the body temperature. Shivering uses a great deal of energy, so the hypothalamus sends out messages of hunger at the same time.

If the need for food and warmth is not met, shivering will eventually consume the body's stored energy. **Hypothermia**, a severe lowering of the body's core temperature, may set in. This can become a life-threatening situation, not only because the body temperature is dropping, but also because the person's control of bodily functions is reduced and the person thinks less rationally.

The hypothalamus also takes action to reduce heat loss when the body temperature begins to drop. Normally, a considerable amount of heat leaves the surface of the body from the blood vessels underlying the skin. The muscles around the arterioles leading to the skin can be constricted to reduce this effect. (When this happens the skin appears to turn from pink to white.) Under these circumstances it is easy for frostbite to occur. Frozen skin must be very carefully warmed, preferably under medical supervision, if it is not to be damaged.

Response to High Temperatures

The hypothalamus also takes immediate action if the blood temperature goes too high. In this situation, the arterioles to the skin open, permitting more blood to flow to the surface to release heat. (When this happens the skin appears to turn from pink to red.) The hypothalamus then initiates sweating. The evaporation of water from the skin surface absorbs considerable quantities of heat from the blood, thus lowering its temperature. Messages are also sent to other portions of the brain, directing the person to drink fluids and reduce activity. Ignoring these messages can lead to **heat exhaustion**. Drinking cool nonalcoholic fluids and bathing the body with cool water can help to lower the internal temperature, as you can see in Figure 3.5. However, alcoholic beverages can make the situation worse.

Figure 3.5.
Becoming overheated is dangerous. The body temperature must be lowered and lost fluids replaced.

Fever

Bacterial infections seem to alter the temperature-regulating mechanisms of the body. Part of the action of the white blood cells in combatting infection appears to be a "resetting of the thermostat" a few degrees higher. This produces the effect known as fever. Although the mechanism is still under investigation, research is indicating that fever is beneficial in fighting infection, so steps should not be taken to reduce it unless it is severe.

Regulation of Blood Glucose

Another important feedback system regulates the level of glucose in the blood. The digestion of each meal dumps large amounts of glucose into the circulatory system. This glucose must be quickly converted to glycogen or fat for storage, or it will be excreted when it reaches the kidney. The homeostatic mechanism for this process is shown in Figure 3.6. The pancreas (Figure 3.7) performs most of the monitoring and control for this system. The pancreas is a complex organ consisting of two types of glands. The first are **exocrine** glands, which produce enzymes that enter the digestive system. The other type are **endocrine** glands, which produce hormones that enter the blood stream. The hormones are produced by three types of cells which cluster together in groups called the **Islets of Langerhans**.

Insulin, one of the pancreatic hormones, acts slowly but strongly to lower the amount of glucose in the blood. It accomplishes this by increasing the amount of glucose absorbed by muscle and fat cells, and the amount of glucose converted to glycogen by the liver. **Glucagon**, another pancreatic hormone, acts more rapidly and raises the amount of blood glucose. Glucagon thus works in opposition to insulin, and is said to be *antagonistic* to it. Glucagon is less

Figure 3.6.
Control of Blood Glucose Levels. The cells of the pancreas monitor and regulate the amount of glucose in the blood by secreting various hormones. The liver, muscles, and other cells can store and release glucose as required. People with hormone imbalances must discipline their carbohydrate intake to assist in this regulatory process.

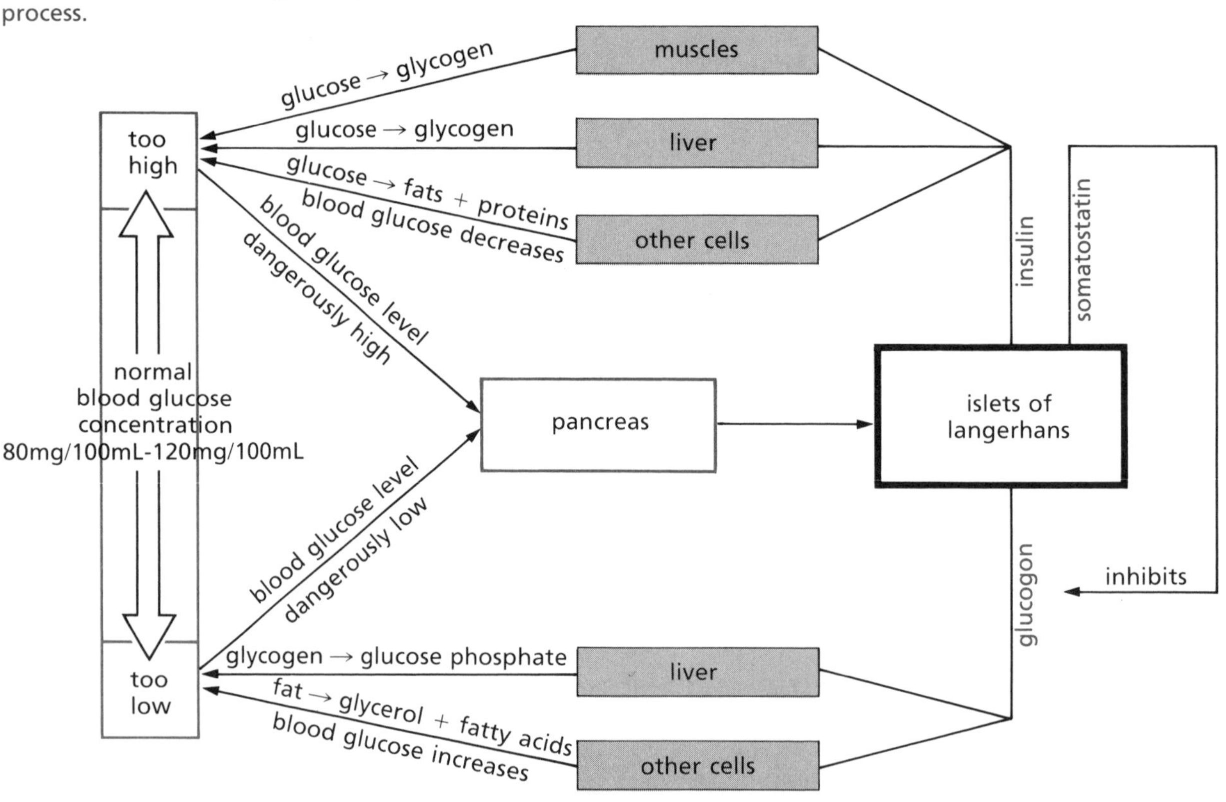

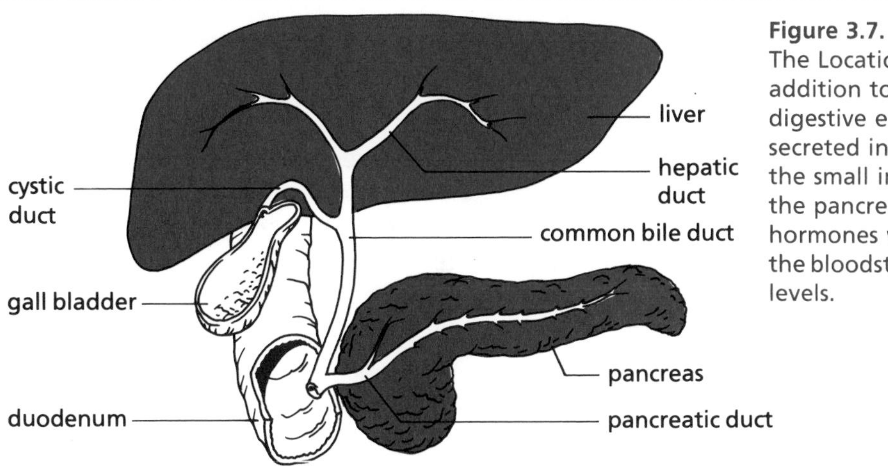

Figure 3.7.
The Location of the Pancreas. In addition to its production of digestive enzymes which are secreted into the upper part of the small intestine (duodenum), the pancreas also produces hormones which are released into the bloodstream to control glucose levels.

powerful than insulin. This arrangement has distinct advantages. Because insulin is slow-acting, variations in blood glucose tend to cause the release of more insulin than is necessary to stabilize the system. When this occurs, the faster-acting glucagon is released to restore the balance.

If the blood sugar level is high, a third pancreatic hormone, **somatostatin**, is also released. Its function is to inhibit the action of glucagon, thus preventing the release of more glucose. Together, these hormones control the reactions which convert glucose to glycogen, fats, or proteins and release glucose from those substances.

If the blood glucose level remains high, the kidney will remove the excess glucose from the blood and excrete it. However, this action effectively starves the cells and also interferes with the kidneys' ability to reabsorb water. The resulting condition, known as **diabetes mellitus**, can cause death if untreated. The pattern of treatment necessary depends upon whether the condition is caused by a failure of the pancreas to produce sufficient insulin, or by an inability of the cells to respond to insulin.

When the system regulating blood glucose levels fluctuates to the extremes of the normal range, it produces the "highs" and "lows" many of us experience during the course of a day. These variations in blood glucose can usually be controlled by modifying the diet so that it contains less sugar. A diet which is consistently high in sugars and other simple carbohydrates places a very high demand on the pancreas. It now appears that this type of diet is responsible for the increasing development of diabetes in adults.

Regulation of Digestion

The release of digestive enzymes is, again, controlled by feedback systems involving hormones. If these enzymes were released into the digestive organs when no food was present, the cells of those organs could be digested instead. The smell and taste of food initiates the release of digestive enzymes, but, fortunately, this release soon ceases unless food actually enters the stomach. Some of the stomach cells release a hormone, **gastrin**, which travels through the bloodstream back to the stomach to trigger the cells which release gastric juice. This hormone controls about 75% of the secretion and ensures that the amount of enzyme released will be balanced with the amount of food consumed. The release of the remaining 25% of the gastric secretion is under nervous control.

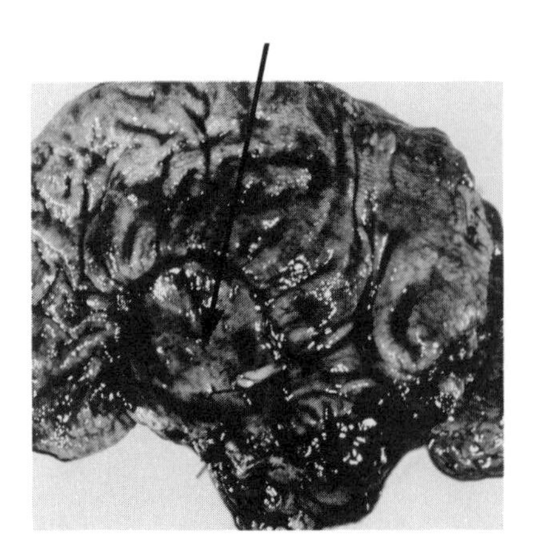

Figure 3.8.
A Photo of a Stomach Ulcer. The presence of digestive enzymes when there is no food for them to act on can damage or destroy the cells lining the digestive system.

Caffeine and alcohol strongly stimulate the release of gastrin. If they are consumed under stress conditions or when little or no food is eaten, too much gastric juice may be released. If this occurs frequently, the result may be gastroenteritis (an inflammation of the lining of the stomach) or an ulcer (destruction of stomach cells), as Figure 3.8 illustrates.

A similar system controls the release of enzymes into the small intestine. **Secretin** and **cholecystokinin** (CCK) are released by the cells of the duodenum (the first part of the small intestine) into the bloodstream when food passes from the stomach into the duodenum. Secretin controls the release of sodium hydrogen carbonate; CCK controls the release of digestive enzymes and bile.

Regulation of Breathing

Feedback systems also control the rate and depth of breathing. Cells can survive for only a few minutes without oxygen. Brain cells are among the most sensitive to oxygen deprivation, so the brain is not the ideal location for a monitoring system. The concentration of both oxygen and carbon dioxide in the blood is monitored in the aorta and the carotid artery before the blood reaches the brain. Figure 3.9 shows the location of these blood vessels. The message is sent to the **medulla oblongata**, a control center embedded deep in

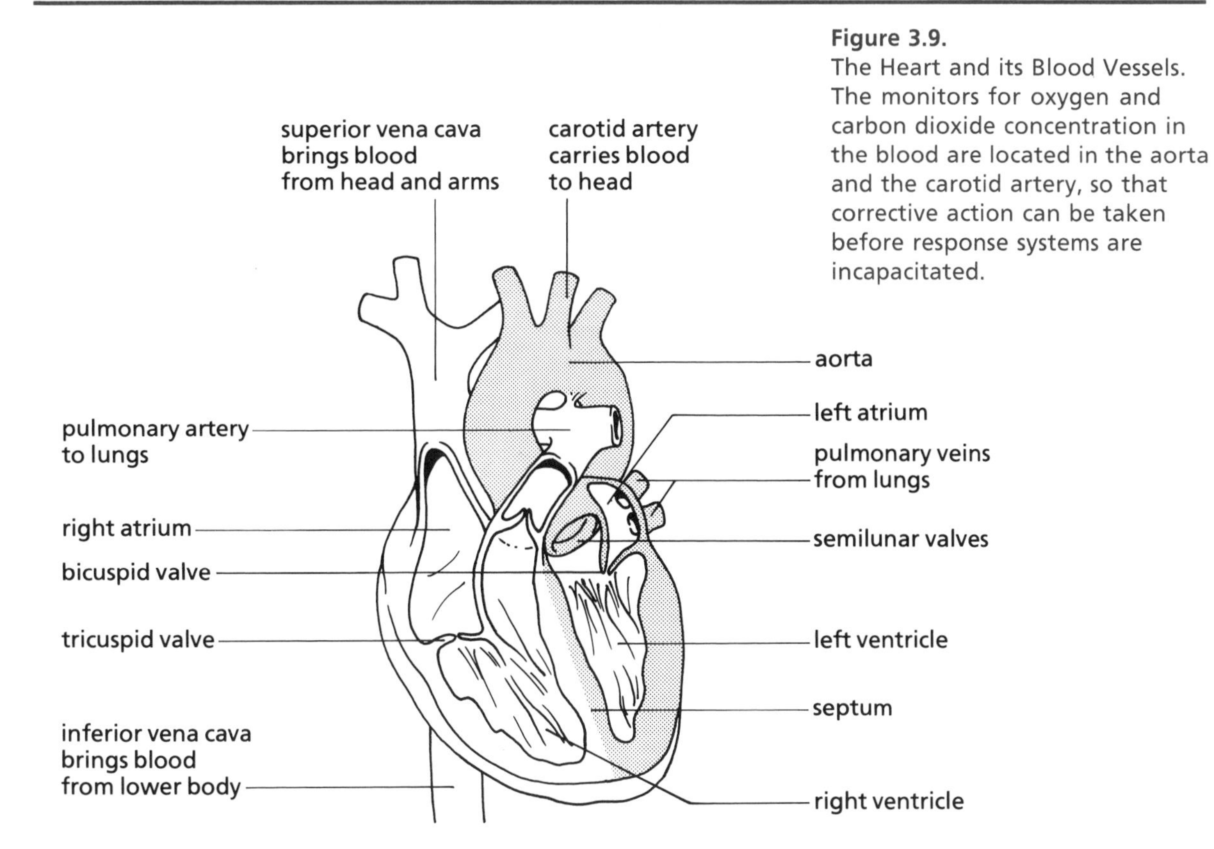

Figure 3.9.
The Heart and its Blood Vessels. The monitors for oxygen and carbon dioxide concentration in the blood are located in the aorta and the carotid artery, so that corrective action can be taken before response systems are incapacitated.

the most protected part of the brain. (The location of the medulla oblongata was shown in Figure 3.3.) Since nerve impulses travel more rapidly than blood, the brain has an opportunity to respond before it is deprived of oxygen. Nerve impulses travel from the medulla oblongata to the diaphragm and the respiratory muscles, signalling the necessary changes in the rate and depth of breathing. These processes are shown in Figure 3.10 (on the next page).

Under normal conditions, it is the carbon dioxide concentration which is responsible for changes in the depth of breathing. As carbon dioxide accumulates in the blood, the blood pH decreases. This increase in acidity causes the muscles of the bronchiole walls to relax. This in turn enlarges the passageways, allowing more gas exchange. You can see these passageways in Figure 3.11 (on the next page).

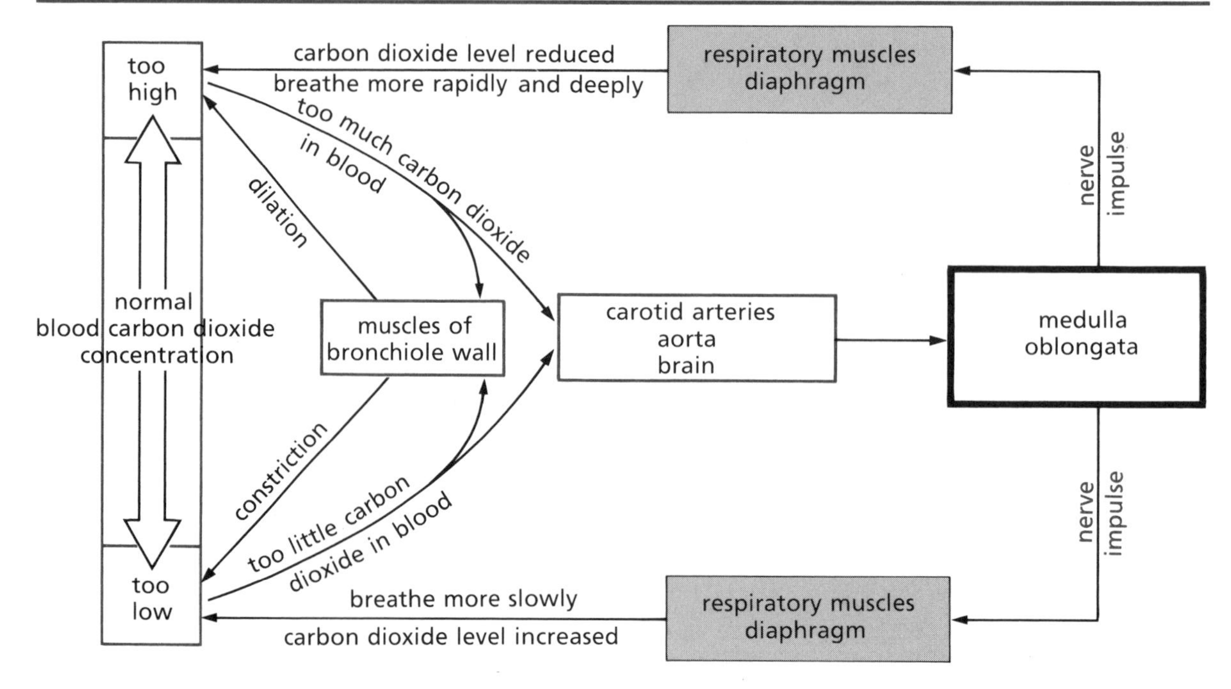

Figure 3.10.
Control of Gas Exchange. The monitors here are within the lungs and the blood vessels leaving the heart. The brain must take action before blood deficient in oxygen impairs its function.

The Effect of High Altitudes

If the oxygen deficiency results from a low oxygen concentration in the air, as, for example, at high altitudes, a long-term adaptation may occur. Since oxygen is transported by the hemoglobin in red blood cells, the body begins to manufacture additional red blood cells, enabling it to transport adequate oxygen. This response, called **acclimation**, takes about three weeks to complete. Athletes who must compete in events held at high altitude, such as the Olympic Games which were held in Mexico City, must arrive at least three weeks before their event to give their bodies time to adjust. They could, of course, train at another location of similar altitude. Few new records are set at athletic events held at altitudes above 1000 m. Some athletes train regularly at high altitudes, feeling that it gives them an advantage at meets held at lower altitudes, particularly in events requiring endurance. People who spend their entire lives at high altitudes have very efficient gas exchange and oxygen transport systems.

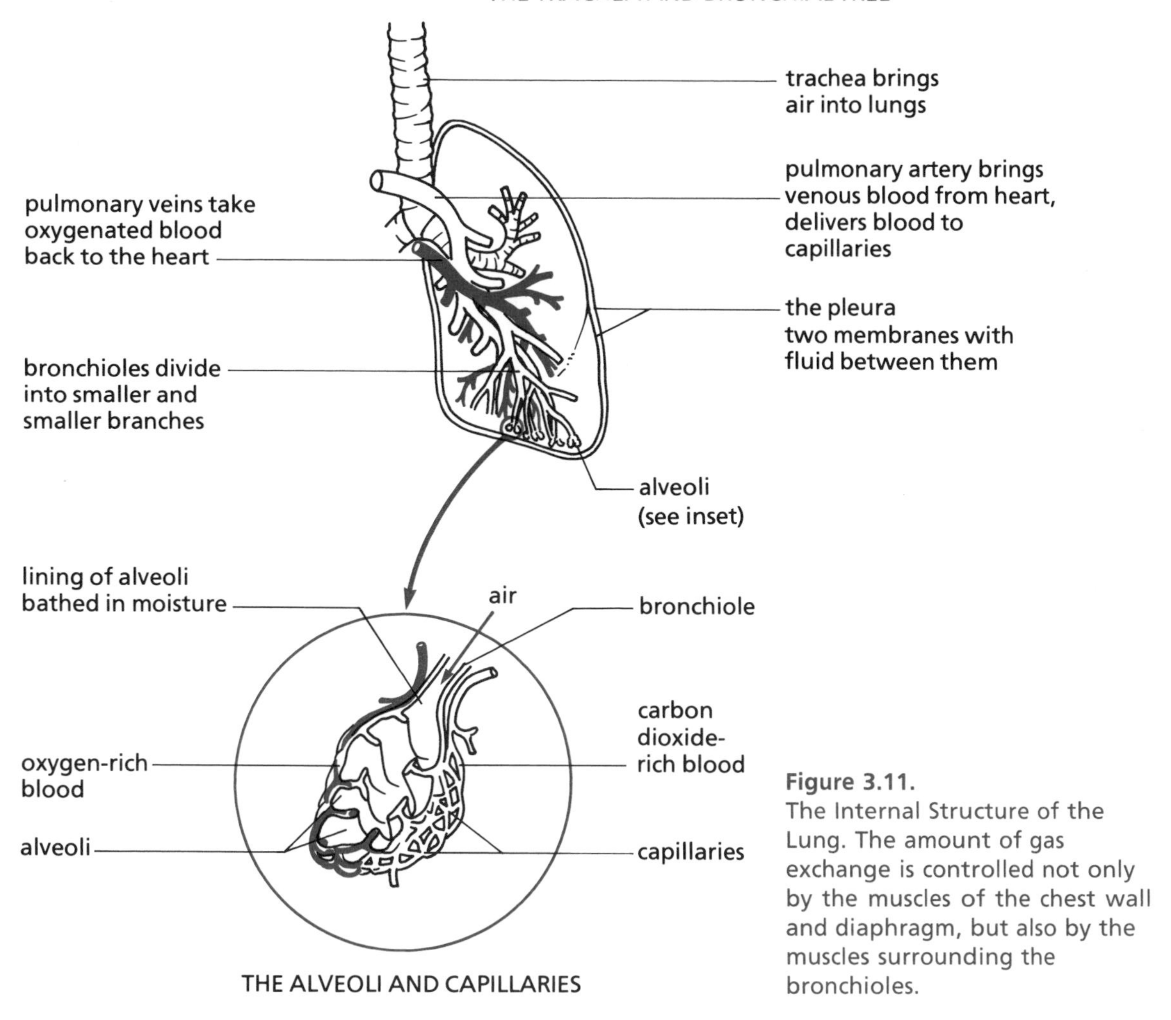

Figure 3.11.
The Internal Structure of the Lung. The amount of gas exchange is controlled not only by the muscles of the chest wall and diaphragm, but also by the muscles surrounding the bronchioles.

Regulation of Other Systems

Both the concentration of ions in the blood and blood pressure are controlled by a series of interlocking feedback systems. Hormones from many endocrine glands are involved in controlling the release and reabsorption of substances by the kidney. The artificial kidney (dialysis equipment) duplicates only the most essential of these processes. A kidney transplant is therefore still a better long-term solution for kidney disfunction.

The hormones of the menstrual cycle form another feedback system — one which provides two very different responses depending upon whether fertilization has occurred.

All the body's chemical processes are controlled by elaborate networks of feedback systems. Interference with any one of these networks tends to affect all the rest.

Upsetting the Balance

Pesticides

Figure 3.12.
Do you use pesticides safely? All pesticides are poisons, and the instructions for their use must be followed accurately.

Many pesticides, such as the one being sprayed in Figure 3.12, function by interfering with the feedback systems of the target pest (often insects). Unfortunately, many of these feedback systems are common to a wide variety of other organisms, among them wild mammals, birds, and fish, pets, and human beings. All these organisms can be harmed by pest control efforts. The process of interference with feedback systems is called **competitive inhibition**. The name arises because the pesticide competes with an enzyme or hormone in one or more systems and prevents it from acting.

If you must use such chemicals, follow the instructions with great care. The most toxic pesticides are not approved for household use; they may be used only by licensed business enterprises which have been fully instructed in their use. If you ever use organophosphate pesticides, be sure to use the proper safety equipment, including breathing apparatus and skin protection such as that illustrated in Figure 3.13. These pesticides are closely related to the chemicals proposed as biological weapons. They act by inhibiting the enzyme **acetylcholinesterase** and thereby interfering with the normal transmission of nerve impulses. Carelessness in their use can result in illness, paralysis, or even death.

There is also mounting evidence that less toxic, fat-soluble pesticides can accumulate in biological food chains to the point where their concentration in predator species such as hawks, wolves, tuna, herons, and even humans can interfere with feedback systems. Nursing mothers must be

Figure 3.13.
The Safe Use of Agricultural Pesticides. These powerful chemicals should be used carefully. Here a greenskeeper, licensed to apply pesticides, is using Paraquat in a sand trap area of a golf course. Note the disposable coveralls and gloves, rubber boots and respirator.

especially careful in using pesticides, since they can be transmitted to infants by the fat in breast milk.

Heavy Metals

Heavy metals such as mercury, cadmium, lead, copper, and zinc tend to combine with the protein molecules in living organisms. Many enzymes function by forming a specific metal/protein complex. The "wrong" metals can interfere with the formation of such complexes, and impair or stop enzyme action. The body cannot break down these abnormal molecules; therefore, the metals cannot be excreted. As a result, very small amounts of these heavy metals in the environment can accumulate in living organisms to produce major effects.

Lead and mercury have a tendency to combine with enzymes in the central nervous system, producing neurological damage. These effects have been observed for a long time. For instance, the expression "mad as a hatter" and the Mad Hatter in *Alice in Wonderland* reflect the fact that hatters used mercury to treat the surfaces of felt hats. There is still cause for concern about both lead and mercury poisoning.

Lead is used in many industrial processes, and the wastes are not always correctly handled. It is also found in some paints and ceramic glazes. Our laws prevent the use of such substances on objects intended to contain food or beverages or on objects which young children might chew. However, this is not the case in all countries. Tourists should be wary of using ceramic souvenirs to serve food or beverages or of purchasing painted toys for children.

The major source of lead pollution today, however, is the leaded gasoline burned in automobiles. The oil industry has been asked to reduce and eventually eliminate lead additives. Young children are particularly susceptible to the effects of lead poisoning. Some parent groups are agitating for the immediate prohibition of lead additives in gasoline. Figure 3.14 shows how you, as a consumer, have an active role in controlling lead pollution.

Figure 3.14.
The choice is yours. Do you use unleaded gasoline or another unleaded fuel such as propane?

The effects of mercury are also well known. The careless disposal of waste mercury compounds by the pulp and paper industry has poisoned many of our lakes and rivers. These compounds result in the formation of methyl mercury, which accumulates in food chains. The effects of eating fish from such waters were first observed in Japan, where they became called *Minamata disease*. Methyl mercury attacks the central nervous system, causing lack of muscle control, convulsions, and even death. These effects were particularly devastating at Minamata, because the mercury-bearing wastes were dumped directly into the bay from which residents obtained the fish that formed a major portion of their diet. Because of similar problems, the eating of fish caught in many areas of Canada is restricted or prohibited entirely.

Mercury is also known to interfere with a protein involved in DNA function, producing birth defects. Pulp and paper mills are no longer permitted to dump mercury-bearing wastes into our waterways. Special funds and tax incentives have been provided to bring the waste management processes of existing mills up to current standards. But we will continue to have a problem for some time with the wastes dumped in the past, since there is presently no known method which is effective in removing them.

Other heavy metals, too, are used by many industries in a variety of processes. Artists are particularly at risk, because many of the pigments they use are based on heavy metals, which provide very pure and stable colours. Artists who work in studios are not protected by industrial health laws, and must take responsibility for the safe use of these substances.

Another major source of heavy metal pollution today is the incineration of wastes. Many people are particularly concerned about the disposal of the batteries which are increasingly being used in portable electronic devices. Most of these batteries are composed of heavy metals. Have you ever noticed the disposal instructions printed on the rechargeable batteries for radios, tape players, and calculators, shown in Figure 3.15?

Researchers are now examining evidence that suggests that combinations of these metals are even more dangerous than the individual metals. For example, copper and zinc in combination are known to be ten times as toxic to fish as either element by itself.

Figure 3.15.
What happens to your discarded batteries?

Possible Solutions

As metals become increasingly more expensive, waste management programs are becoming more efficient in recycling them. Not only is pollution reduced, but more economical use is made of materials.

The developing embryo is particularly sensitive to the effects of competitive inhibition. Pregnant women should therefore avoid eating fish or shellfish from waters known to be polluted with pesticides or industrial chemical wastes.

Our laws give employed pregnant women the right to request transfer to alternative tasks where they are less likely to be exposed to toxic substances.

Increased public awareness and concern are leading to more careful disposal of wastes and improvements in the treatment of waste water and drinking water. Pesticide residues on food substances are now regulated. We have, however, a long way to go before we can assume that all water is safe to drink and all food safe to eat. Each of us should be alert to sources of pesticide and heavy metal pollution and support actions to reduce the hazard for us all.

QUESTIONS FOR REVIEW

SOME WORDS TO KNOW

Match each description given in the left-hand column with a word shown in the right-hand column. DO NOT WRITE IN THIS BOOK.

1. A process which maintains the body in internal balance.
2. The portion of the brain controlling body temperature.
3. A hormone produced by the thyroid gland.
4. A process that will increase body temperature.
5. A condition in which the internal body temperature is too low.
6. A general body response to infection.
7. A gland that secretes into the bloodstream.
8. The hormone which reduces the level of blood glucose.
9. A disease involving inability to regulate blood glucose.
10. A hormone which causes the release of enzymes into the stomach.
11. A hormone which causes the release of enzymes into the small intestine.
12. The portion of the brain which controls breathing.
13. An organ which controls the concentration of ions in the blood.
14. Interference with a process caused by a poison replacing the normal substance.

A. hypothermia
B. medulla oblongata
C. insulin
D. homeostasis
E. CCK
F. thyroxin
G. gastrin
H. hypothalamus
I. endocrine
J. kidney
K. competitive inhibition
L. diabetes mellitus
M. fever
N. shivering

SOME FACTS TO KNOW

1. Why does the body need homeostatic mechanisms?
2. What are the essential components of feedback systems?
3. How are messages relayed through feedback systems?
4. What are the components of the system controlling metabolic rate?
5. Describe the various adjustments that can regulate body temperature.
6. What happens when the body cannot control its internal temperature?
7. Why are the cells of the Islets of Langerhans so important?
8. What happens when the body cannot regulate the blood glucose level?
9. Why must the release of digestive enzymes be under hormone control?
10. How are the rate and depth of breathing controlled?
11. How do many pesticides accomplish their task?
12. What effect can heavy metals have on feedback systems?

QUESTIONS FOR RESEARCH

1. Some people must live and work in the extremely cold conditions of the Arctic. What changes in your normal lifestyle would you expect to have to make if you were to do this? Look up the results of some of the research conducted by the Armed Forces on this subject. Evaluate the traditional lifestyle of the Inuit from this viewpoint.
2. Medical science has learned much about treating patients whose body temperature has dropped very low, often enabling them to make a full recovery. What are the currently recommended procedures in such a situation?
3. For some operations, doctors artificially induce a condition similar to hypothermia. What would be the advantages of such a treatment?
4. Some alcoholic beverages contain substances which inhibit the reabsorption of water by the kidney. What would be the effects of consuming a large quantity of one of these beverages?
5. Many office buildings have sealed windows, to allow control of ventilation. If the level of gas exchange is too low, however, the concentration of carbon dioxide in the circulating air may increase during the day. What would be the effect on the people working in the building?

6. Modern mountain climbers and pilots frequently use supplementary oxygen at high altitudes. What sometimes happened to early climbers and aviators without such equipment?

7. Athletic events held at high altitudes, such as the Olympic Games held in Mexico City, are a severe test for the competitors. Yet athletes who habitually train at such altitudes claim that it gives them an advantage. What is the basis for this claim? How much of an advantage might such training provide? What current practices are based on this idea?

8. Find out more about the two feedback systems mentioned only briefly in this chapter: the filtration system of the kidney and the menstrual cycle. Refer to other biology texts, and draw diagrams to assist you in understanding these systems.

9. Investigate pesticide use in your area. Report on one pesticide presently in use, making certain your report includes answers to the following questions: How is it applied? How does it act? What is its risk to residents? How persistent is it (how long does it last?). Does it remain in the soil or on crops? Does it enter the food chain? Can it accumulate? How necessary is it? Are there safer pesticides available? Are there alternative means of controlling the specific pest or pests? Based on the evidence you have gathered, do you think its use should be continued? What might be the consequences of discontinuing its use? How might you have influence on its use?

10. What heavy metals are likely to be found in industrial wastes in your area? Select one heavy metal and report on it. How is it used by the industry? Are there safer alternatives? Can this substance be recycled? If so, is it? If not, is the possibility of recycling being researched? How are its wastes treated? What is the method of disposal? Can this substance contaminate the ground water? Is this being monitored? How safe is your water supply from heavy metal contamination?

11. Are chemical pollutants other than pesticides and heavy metals a problem in your area? If so report on them, including answers to the following questions: Who monitors the disposal of toxic chemicals from industry? Who monitors the safety of the water you drink and the air you breathe? How effective is this process? What happens when toxic chemicals are spilled or deposited in unsuitable places? What makes a place "unsuitable?" How might you become involved?

Chapter 4

The Manufacture of Proteins

- Protein Structure
- The Importance of Proteins
- The Structure of DNA
- Protein Synthesis
- Genetic Mutations
- Recombinant DNA

Chapter 4

The Manufacture of Proteins

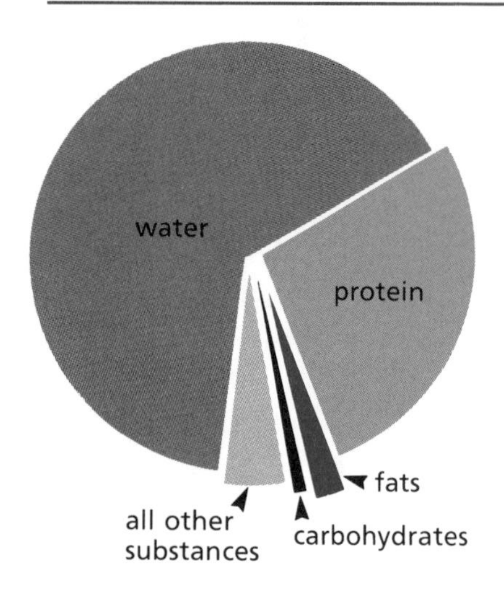

Figure 4.1.
The Composition of a Typical Cell. Note the proportion of the cell that is protein.

Consider the people you know. There is little difficulty in identifying each individual. With the exception of identical twins, each person has physical characteristics, such as eye and hair colour and facial proportions, that are distinctly different from those of others. People also differ in many less visible ways. Yet all people are made of cells and all cells are made of the same chemicals. How, then, can each person be so different?

Seventy-five percent of the typical cell is composed of water, as you can see in Figure 4.1. Since this portion clearly cannot account for the differences you observe, the answer must lie in the 25% of the cell that is *not* water. Of this, 20% is protein, while carbohydrates, fats, vitamins, and all other substances of the cell combined form the remaining 5%. Proteins are far more complex molecules than the others; therefore, they might be expected to provide the answer — as, indeed, they do.

Protein Structure

Proteins are complex networks of amino acids. The structure of insulin, a relatively simple protein, is shown in Figure 4.2. It has been found that 20 different amino acids can be used as building blocks for proteins. Hundreds or even thousands of these 20 amino acids may be linked together in different ways to form an almost limitless variety of proteins. The proteins may form simple chains, or they may be folded, twisted, or cross-linked. In fact, proteins can be constructed to possess whatever characteristics the cell requires to perform its function.

The proteins that perform a specific function have a similar structure, regardless of the organism in which they are found. The proteins of closely related organisms have fewer amino acid differences than do the proteins of less closely related organisms. Thus, it is possible in many cases to extract proteins, such as insulin, from other mammals and use them to treat human deficiencies. Extracts from non-mammals would be more likely to cause an allergic reaction, since more amino acid differences would exist. Antibodies, too, can be extracted from other mammals and used to immunize humans. But these are not human proteins, and human cells can recognize that fact. As a result, the recipient's white cells attempt to remove the invading protein. The intensity of this rejection reaction depends on the source of the protein and the body chemistry of the recipient.

Figure 4.2.
The Structure of Insulin – A Simple Protein. The two amino acid chains are cross-linked by sulphur atoms.

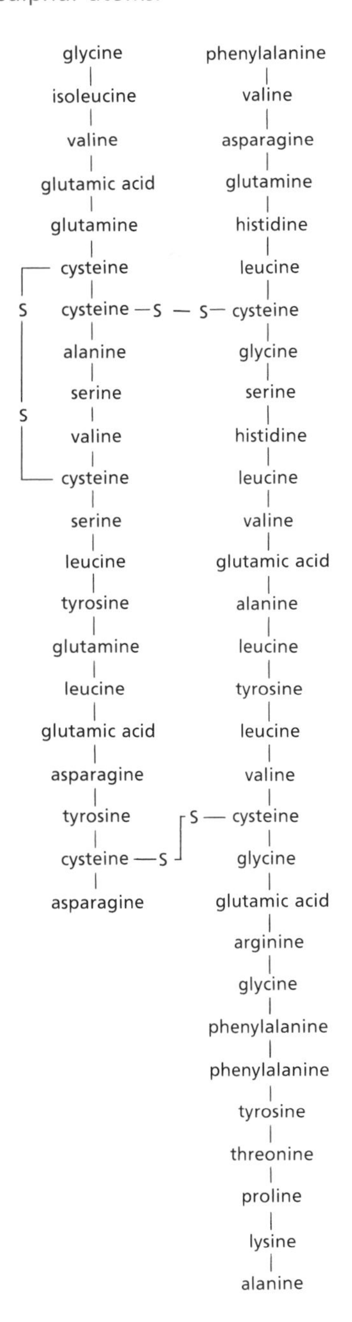

The Importance of Proteins

How can a cell distinguish between "self" and "other"? The answer lies in the proteins on the surface of the cell membrane. A small portion of the membrane is shown in Figure 4.3 (on the next page). These proteins are unique from individual to individual. The more closely related the individuals are, the more similar will be the proteins on the surfaces of their cells.

It is these proteins that are compared when blood is typed before a blood transfusion. Thus, all individuals of blood type A have blood cell surface proteins which are similar enough that there will not be a severe reaction when they are mixed. But another aspect of these proteins must also be compared. Before a transfusion, the blood must also be typed for **Rh factor**, another protein which may be found on the cell surface. If the samples still coincide, they are then *cross-matched* (mixed) to ensure that other surface protein characteristics are also compatible.

Tissue typing is carried out to match potential recipients with organs that are available for transplant. It has a basis similar to blood typing. Unless the cell proteins of the donated organ are similar to those of the recipient, the organ will be rejected by the recipient. This increases the stress on an already ailing system. The ideal donor would, of course, be an identical twin, but this situation is rare. Close

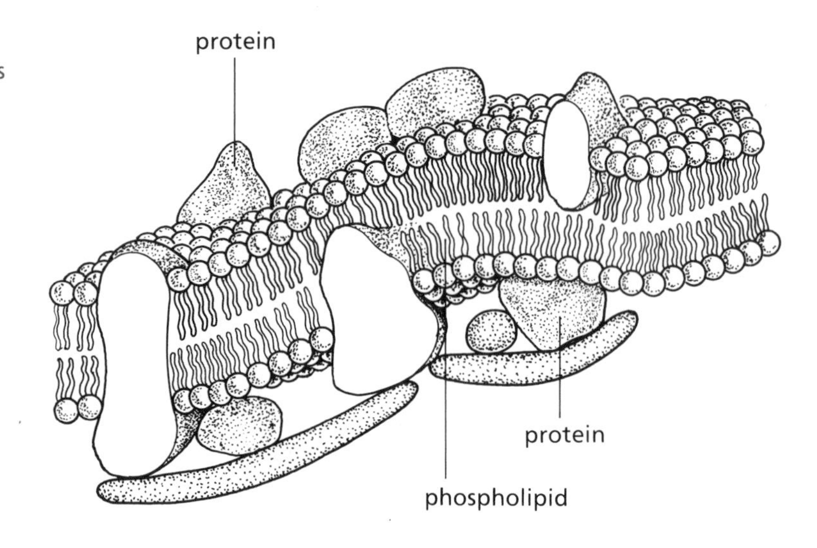

Figure 4.3.
The Structure of the Cell Membrane. Some of the proteins are on the surface, while others penetrate the membrane. The structure of these proteins is characteristic of each individual human being.

relatives may have proteins that are sufficiently similar to be tolerated by the recipient. However, a stranger living in another city or country might also have compatible proteins. In recognition of this possibility, a computer network has been established, linking hospitals across North America. Today, an organ from an accident victim can be matched to the most appropriate recipient, regardless of location. The more organ donors there are, the greater the possibility that a suitable organ will be found in time to save the recipient's life. When a match is made, nutrients are supplied to the organ to keep it healthy during transport to the recipient.

Why do identical twins have almost identical proteins? Why do children have proteins that are so similar to those of their parents and grandparents that this characteristic can be used to help establish (or disprove) paternity? To find the answer, we must consider the mechanism of heredity.

Chromosomes, found in the nucleus of the cell, are the parts of the cell which carry hereditary information. The DNA (**deoxyribonucleic acid**) molecules which make up the chromosomes transfer the information that determines the composition of proteins. A **gene** is a portion of a DNA molecule which contains a set of instructions for the manufacture of a specific protein.

The Structure of DNA

To understand how the body constructs proteins, it is necessary to know the basic structure of DNA. Although DNA was identified in 1869, its importance was not recognized for many years. Serious efforts to work out its structure began after the Second World War. In 1953, James Watson and Francis Crick were successful in working out its complex arrangement of smaller units.

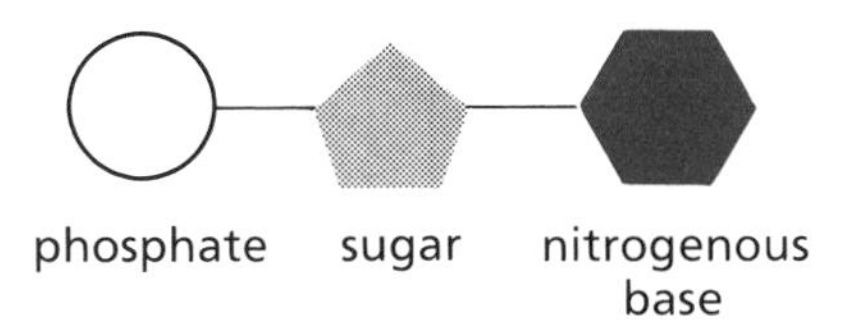

Figure 4.4.
A Typical Nucleotide. Four different nucleotides compose DNA. They each contain a sugar, a phosphate, and one of the nitrogenous bases: adenine, thymine, guanine, or cytosine.

DNA is the largest molecule in living organisms, but like most large biological molecules it is composed of much simpler units. There are three basic units that link together to form the ladder-like DNA molecule. Phosphate groups and sugars (deoxyribose) form the "sides" of the ladder, while larger molecules called **nitrogenous bases** link together to form the "rungs". Four different nitrogenous bases are found in DNA: **adenine, thymine, guanine**, and **cytosine**. The combination of one phosphate group, one sugar, and one nitrogenous base is called a **nucleotide**, shown in Figure 4.4. Note that the sugar is always the central unit.

A bond can form between the phosphate group of one nucleotide and the sugar of another nucleotide to form a chain of nucleotides, as Figure 4.5 illustrates.

Watson and Crick recognized that weaker bonds may also form between the nitrogenous bases of the nucleotides. Because of their structures, however, adenine can only bond with thymine, and cytosine can only bond with guanine. The result is two chains of nucleotides pairing to form the typical ladder-like DNA structure shown in Figure 4.6 (on the next page). Note the relationship between the two strands: where thymine appears on one side, adenine always appears on the other; guanine on one side is always matched by cytosine on the other. Thus, the two strands are said to be *complementary*.

Just as a telephone cord is coiled to reduce the space it occupies, so, too, the DNA ladder is coiled into the **double helix** arrangement shown in Figure 4.7 (on the next page).

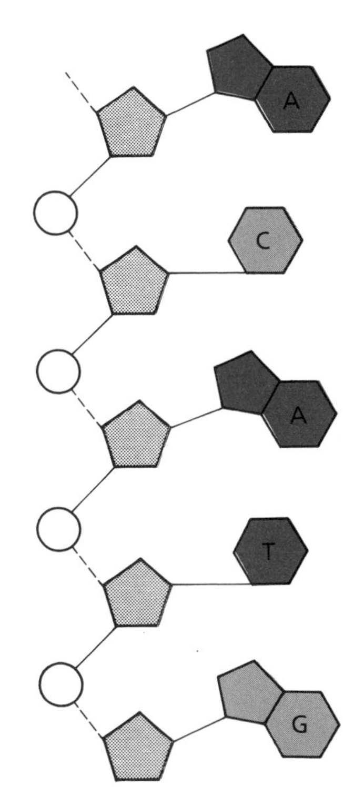

Figure 4.5.
A Chain of Nucleotides. Bonds between the sugar and phosphate groups link the four nucleotides together in any sequence.

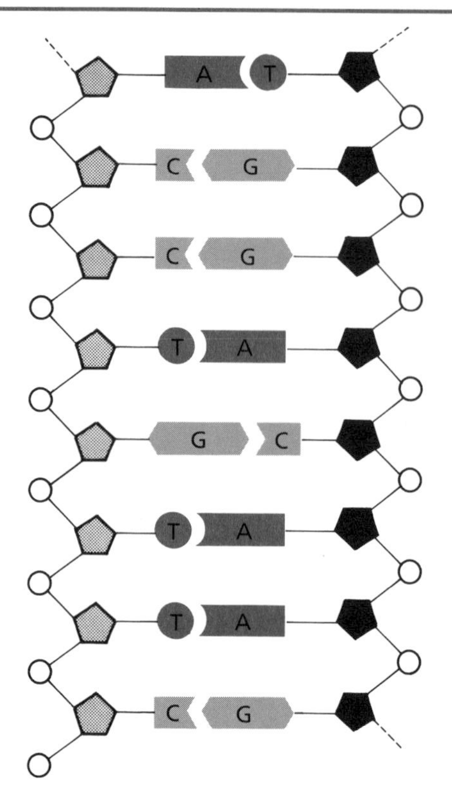

Figure 4.6.
Complementary Strands of Nucleotides. Adenine can only pair with thymine, and guanine with cytosine, but a complex pattern still results.

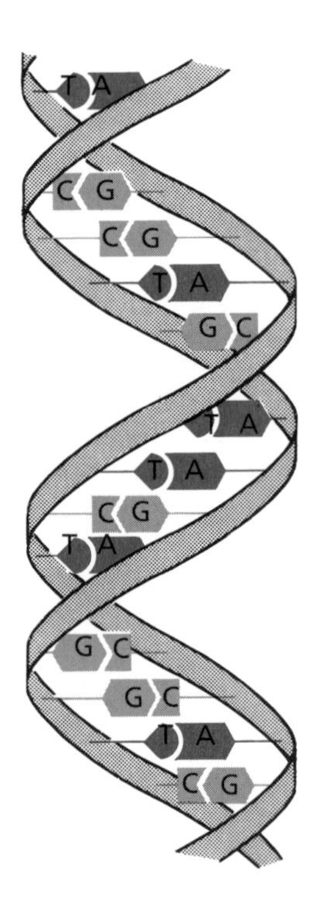

Figure 4.7.
The Double Helix Structure of DNA. This structure may stretch or contract, depending upon the activity of the cell.

The DNA of a single chromosome may contain six million turns and measure as much as 2.2 cm long when fully extended. The double helix shape enables the chromosomes to be tightly coiled for transfer during cell division. When the information carried by DNA is needed by the cell, the required portion of the chromosome is stretched out and partially uncoiled.

DNA Replication

When a new cell is formed, it requires a complete set of DNA to carry out all its activities. The first stage of cell duplication therefore must be the formation of a replica of the DNA. Figure 4.8 shows how the complementary structure of the two strands makes this possible. As the DNA uncoils and separates, each nucleotide attracts a new partner nucleotide from those floating in the cytoplasm of the cell. This process rebuilds two ladder-like structures, each identical to the original. The right half acts as a template for reconstructing a new left half, while the left half assembles a new right half.

The details of this process are still being investigated, but it appears to involve several chemical reactions and at least seven enzymes. The result is two identical DNA molecules. Each chromosome forms a replica of itself in this way so that the new cell has a complete set of genetic information. This replication process appears to start at any of some 6000 sites along the chromosome, and is completed in about two minutes. If it began at one end of the chromosome and proceeded to the other, it would take about two weeks to complete!

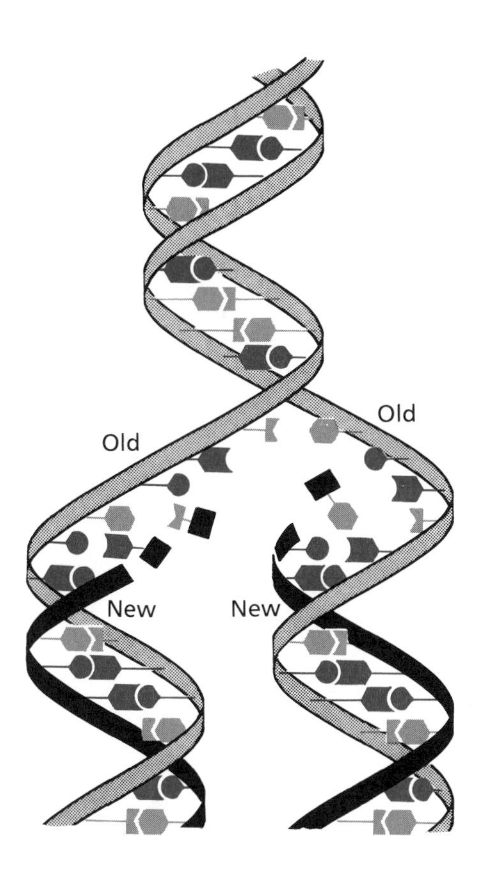

Figure 4.8.
DNA Replication. Notice how each half of the original DNA rebuilds the opposite half to form two matching strands of DNA.

Protein Synthesis

Each chromosome within a cell, and therefore each strand of DNA, carries many genes. Each gene controls the production of a specific protein. Since all enzymes are proteins, this arrangement provides control over all the chemical reactions of the cell. The DNA is too valuable for the cell to use directly for protein synthesis, however. It remains in the nucleus as a "master copy", and "working copies" of the segments required to make the needed proteins are sent out into the surrounding cytoplasm. These smaller segments are called **ribonucleic acid** (RNA).

RNA Transcription

To form RNA, the segment of DNA that forms a single gene is spread apart by enzyme action. One side of the open DNA strand is then used as a base to form the RNA, as in the replication of DNA. But this process, called **transcription**,

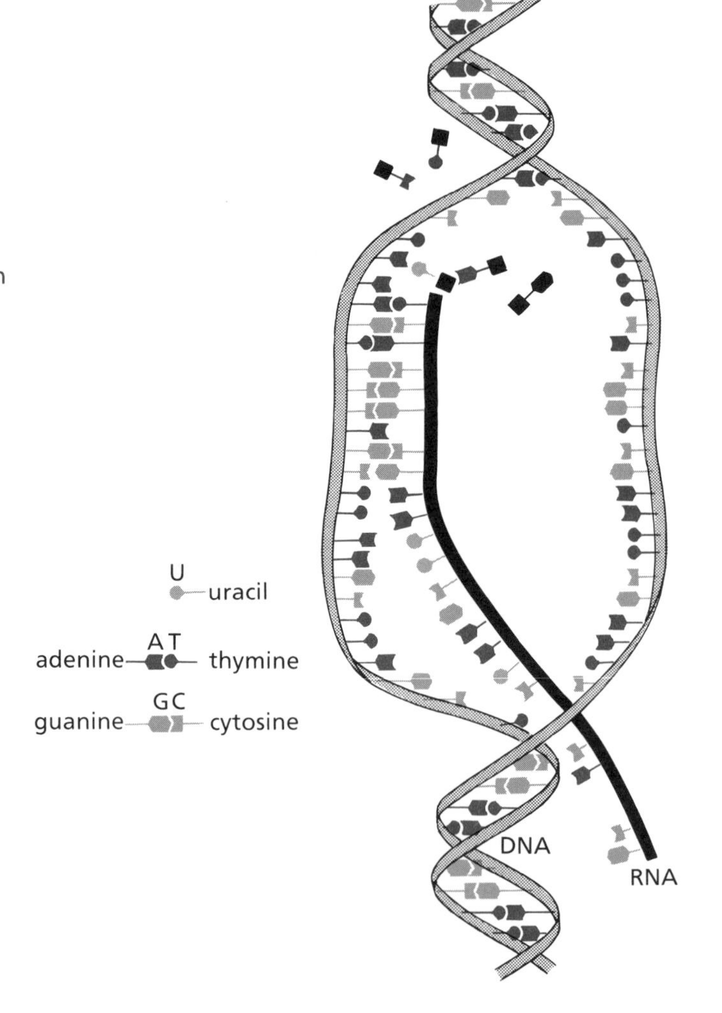

Figure 4.9.
RNA Transcription. The base uracil replaces thymine in the formation of RNA.

differs in two important aspects from DNA replication. First, a different nitrogenous base, **uracil**, replaces thymine in the RNA molecule. Second, the resulting RNA is a single-stranded molecule, not double-stranded like DNA. Figure 4.9 illustrates the transcription of RNA from DNA.

The Types of RNA

Cells contain three different types of RNA, all involved in protein synthesis. Protein synthesis takes place on the surface of the **ribosomes**, cell organelles now known to be clusters of **ribosomal RNA (rRNA)** and proteins. This type of RNA helps in assembling the protein. The second type of RNA codes for a specific protein and is called **messenger RNA (mRNA)**. It bonds to the ribosome to serve as a template for organizing amino acids into a protein. The third type of RNA is called **transfer RNA (tRNA)**. Its function is to collect the required amino acids and place them in position on the mRNA. The three work together as follows: Amino acids from the cytoplasm of the cell are collected by the tRNA, placed into position on the mRNA molecule, and then linked together to form a protein by the action of the rRNA.

The Triplet Code

How, you may wonder, does the mRNA specify which amino acid is required? The answer lies in the sequence of nitrogenous bases that make up mRNA. There are only four different bases composing mRNA: adenine, uracil, cytosine, and guanine. Yet twenty different amino acids could be selected for use in building proteins. Obviously, a single base cannot specify one amino acid. A few calculations quickly show that even using bases in sets of two would not provide enough combinations to identify each amino acid. Researchers soon realized that it requires a sequence of three bases to obtain enough combinations to specify a particular amino acid; therefore, the RNA code is called a **triplet code**.

Each sequence of three bases is referred to as a **codon**. Four bases, selected three at a time, can form 64 different codons. Since there are only 20 different amino acids involved in protein synthesis, some codons must specify the same amino acids. Much painstaking research has tested all the possibilities to determine the amino acid selected by each codon. The results of this effort are shown in Table 4.1. Three of the codons do not select an amino acid; instead, they signal the completion of the series of codons for that protein.

Table 4.1
The Codons Specifying Each Amino Acid. Note that some of the codons for the same amino acid have the first two letters in common.

AMINO ACID	ABBREVIATION	CODONS
Alanine	(Ala)	GCU GCC GCA GCG
Arginine	(Arg)	CGU CGC CGA CGG AGA AGG
Asparagine	(Asn)	AAU AAC
Aspartic Acid	(Asp)	GAU GAC
Cysteine	(Cys)	UGU UGC
Glutamic Acid	(Glu)	GAA GAG
Glutamine	(Gln)	CAA CAG
Glycine	(Gly)	GGU GGC GGA GGG
Histidine	(His)	CAU CAC
Isoleucine	(Ile)	AUU AUC AUA
Leucine	(Leu)	CUU CUC CUA CUG UUA UUG
Lysine	(Lys)	AAA AAG
Methionine	(Met)	AUG
Phenylalanine	(Phe)	UUU UUC
Proline	(Pro)	CCU CCC CCA CCG
Serine	(Ser)	UCU UCC UCA UCG AGU AGC
Threonine	(Thr)	ACU ACC ACA ACG
Tryptophan	(Trp)	UGG
Tyrosine	(Tyr)	UAU UAC
Valine	(Val)	GUU GUC GUA GUG
STOP		UAA UAG UGA

Translation

The mRNA can be considered to be the instructions for assembling amino acids in a specific order. The tRNA folds itself up so that it forms three loops whose free ends come together, as in Figure 4.10. One loop binds to the enzyme which catalyzes the fusion of the amino acid into the developing protein. The opposite loop appears to attach itself temporarily to the ribosome. The free ends pick up the required amino acid. The end loop is called the **anticodon**. It contains bases that are complementary to those of the mRNA codon. The tRNA can therefore plug into the right spot on the mRNA to place the amino acid in the correct sequence in the protein. This process, called **translation**, can be seen in Figure 4.11.

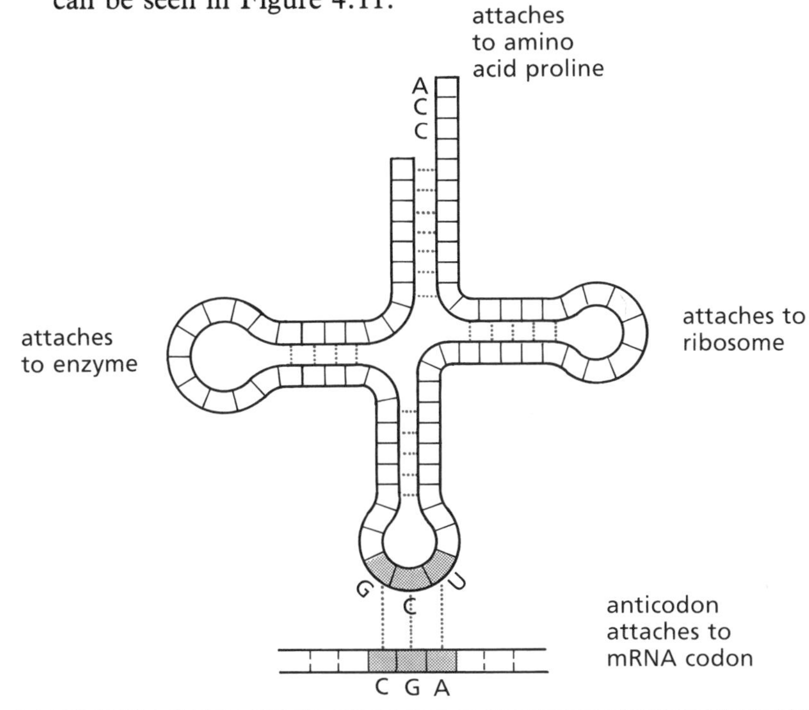

Figure 4.10.
A Transfer RNA Molecule. Each of the lobes has a specific function which enables the amino acid to be correctly linked into the protein.

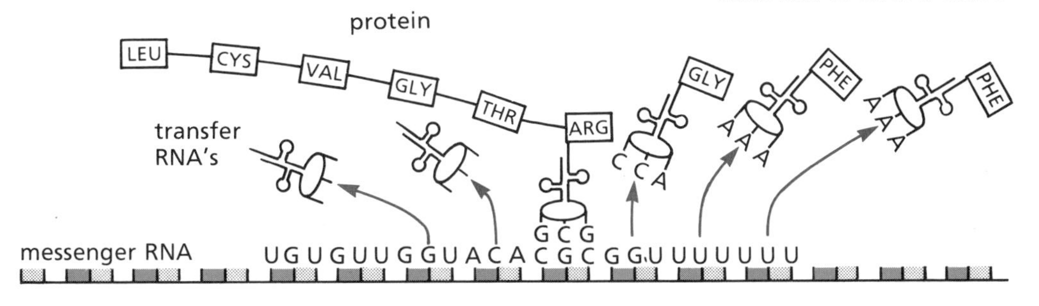

Figure 4.11.
During the process of protein synthesis, the mRNA is attached to one or more ribosomes. Each tRNA attaches its amino acid at the appropriate codon and the amino acids link to form a chain.

The differences between individuals, discussed at the start of this chapter, are the result of this very complex process which determines the kinds and quantities of proteins made by the cells. This capability is inherited. Therefore, the more closely related two individuals are, the more similar their proteins will be. Identical twins, who develop from a single fertilized egg, have very few protein differences. Children will have proteins similar to those of their parents, grandparents, and other relatives. Until recent times, interbreeding was restricted by geographic barriers, so people from one area tend to have proteins similar to each other's, but different from those of people from other regions.

Genetic Mutations

The structure of the DNA molecule is remarkably stable, but changes can occur in it. Any alteration in the normal sequence of bases is referred to as a **mutation**. A number of agents, such as radiation and certain chemicals, are known to induce mutations when they come into contact with DNA. Mutations whose cause is unknown are said to be *spontaneous*. If these changes occur in single body cells or even in a small group of cells, they may be of little consequence. On the other hand, mutations in the reproductive cells will affect every cell of the offspring. For this reason, the reproductive organs are carefully shielded during routine X-rays.

Mutations often involve a change in a single nucleotide of the DNA molecule. This alteration is then copied during RNA transcription. The effect of the mutation on protein synthesis depends on where in the codon the change occurs. If it occurs in one of the first two bases, a different amino acid will be substituted for the correct one. If the change is in the third base, the mutation may be "silent" (that is, not apparent) if the new codon still specifies the same amino acid.

A change in a single amino acid may not seem significant, yet it can produce drastic results. For example, the substitution of adenine for thymine at a specific location in the DNA coding for hemoglobin results in valine replacing glutamic acid at the corresponding location in the hemoglobin produced. This rare and seemingly minor change produces the condition known as **sickle cell anemia**. The blood cells containing the mutated hemoglobin tend to distort into a sickle shape, as in Figure 4.12, when they release their oxygen to the body cells or at low atmospheric pressure. This effect can disrupt capillary circulation. Also, these red blood cells tend to rupture easily, producing severe anemia. Sickle cell anemia is often fatal. Strangely, however, the sickle cell gene provides an ability to resist malaria. For this reason, it may appear in people of African or Mediterranean ancestry, since in those areas it provides a survival advantage.

A number of hereditary diseases are the result of such mutations. In some cases, the missing protein has been identified, and patients can now be treated by being supplied with that missing substance. This is the basis of the treatment for **cystic fibrosis**. Children with this condition do not produce the pancreatic enzymes needed to complete the digestion of food. Pills containing extracts of these enzymes are supplied to the children to enable them to develop normally.

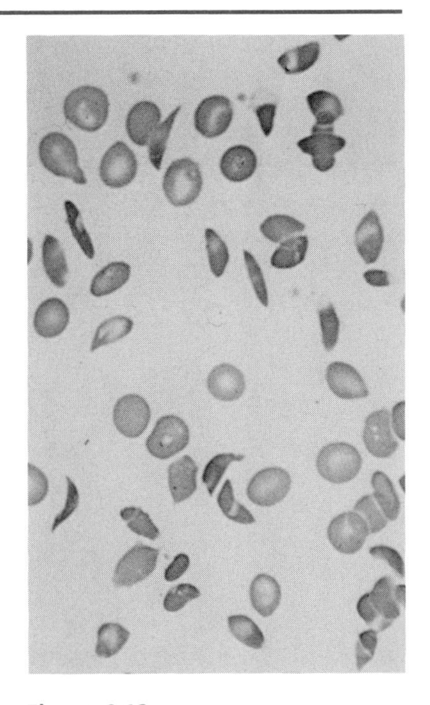

Figure 4.12.
Blood Sample from a Patient with Sickle Cell Anemia. One of the main components of red blood cells is the protein hemoglobin. Normal hemoglobin content results in a disc shaped red blood cell. Sickle cells decrease the oxygen carrying ability of the blood, and tend to gel in the bloodstream.

In other cases, the individual can be kept healthy as long as certain substances are removed from the diet. For example, some babies are born with a condition called **phenylketonuria** (PKU). They are unable to convert excess phenylalanine into tyrosine. As the phenylalanine accumulates in the blood, it causes mental retardation and a tendency towards epilepsy. These children are also very pale, because the tyrosine needed to produce skin pigments is not produced. If such babies are fed a diet from which phenylalanine has been removed, these effects can be prevented. Newborn babies are now routinely tested for PKU.

In many instances, however, the mechanisms of hereditary ailments have not yet been identified. Much research is directed toward the identification of the genes responsible, the mechanism of their action, and the appropriate treatment methods.

Recombinant DNA

Intensive research into the structure and function of DNA and RNA has resulted in the development of a new field of science, **biotechnology**. The first breakthrough was the discovery of a group of enzymes called **restriction endonucleases**. These enzymes can be considered as being tiny shears that will cut DNA strands between specific sequences of bases. Their use has made it much simpler to work out the precise order of the nucleotides in a segment of DNA.

Restriction endonucleases have also made it possible to cut and join different fragments of DNA. Much of this research has been conducted using specially developed mutant strains of *E. coli* bacteria which are incapable of survival outside the laboratory.

Viruses consist largely of DNA. They function and reproduce themselves by injecting their DNA into a cell. They can also be used in the laboratory to introduce DNA into a cell as the researcher requires.

The viruses which attack bacteria cells are called **bacteriophages**. Bacteriophages and the DNA carrying the desired gene are mixed together, and a restriction endonuclease is added. The result is that all the DNA fragments are cut at corresponding locations. Other enzymes, **DNA ligase** and **DNA polymerase**, can then be added to cause the fragments to rejoin. Some of the bacteriophage DNA will now contain the desired DNA, called **recombinant DNA**. This process is illustrated in Figure 4.13. The viruses are allowed to infect *E. coli* bacteria, which are then cultured. The bacteria that show the desired characteristic are isolated, and more are cultured.

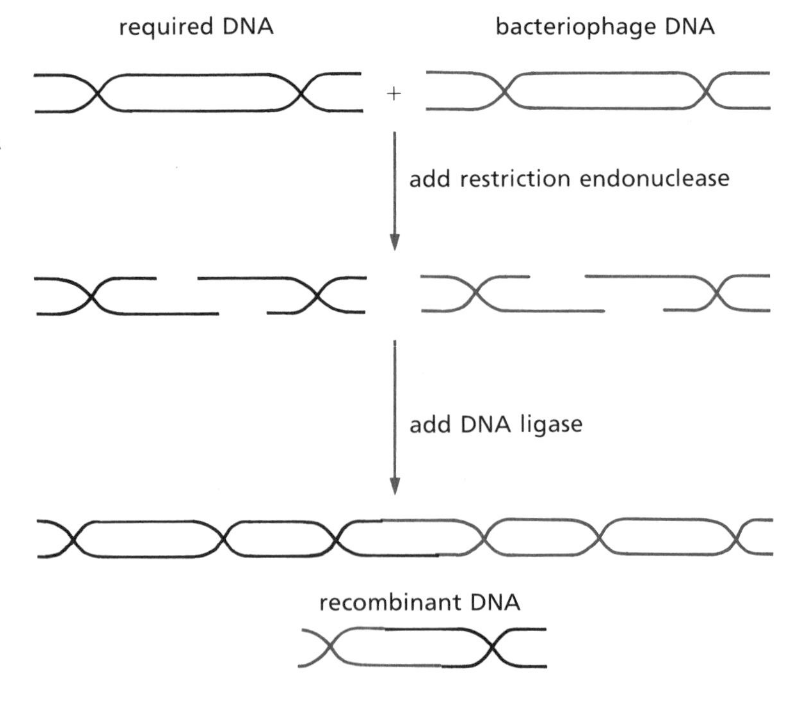

Figure 4.13.
Recombinant DNA. Restriction endonucleases sever the DNA between specific pairs of nitrogenous bases. When the DNA is rejoined by the enzymes DNA ligase and DNA polymerase, some pieces will have interchanged. Thus, some DNA strands will contain different information than they did previously.

Studies of how viruses infect cells led to the discovery of another enzyme, **reverse transcriptase**. This enzyme is able to cause the synthesis of a DNA segment from a strand of mRNA. Thus, it is now possible for researchers to produce an artificial gene containing the DNA with the instructions for a desired protein. Thus, researchers can now produce proteins identical to human proteins. These proteins do not cause the allergic reactions that result from the use of proteins from other mammals. It is possible to produce them in relatively large quantities, although at present most such products are still at the research stage. However, insulin produced in this way already provides relief for those who suffer side-effects from insulin from non-human sources.

The ability to manufacture human growth hormone has offered the hope of normal height to many children who would otherwise never reach adult size. Previously, the limited amount of hormone that could be extracted from donated pituitary glands was severely rationed. The children needing it could be given just enough to permit them to grow to minimum adult height. Now, sufficient growth hormone is available to permit them to attain normal height.

Another substance already being produced is called **clotting factor VIII**. It is of great importance to hemophiliacs. The blood of hemophiliacs does not clot normally; therefore, they require large quantities of this clotting factor before surgery or tooth extraction. Even minor injuries can cause hemorrhage into muscles and joints, requiring treatment with clotting factor. Before synthetic clotting factor was available, hemophiliacs ran the risk of infection from viruses carried with the clotting factor extracted from donated blood.

Vaccines and antibodies have also been synthesized using recombinant DNA techniques. These too, reduce the possibility of side-effects from allergic reactions caused by using another organism to produce the protein. Genetic engineering is even producing vaccines against diseases, such as malaria, for which vaccines have not previously been available. These developments have particular significance for the tropical regions of the world where such diseases are always present.

Many pieces of the DNA puzzle have yet to be put in place. In 1983, an American researcher, Barbara McClintock, received the Nobel prize for her discovery of "jumping genes". These fragments of DNA move from place to place along the chromosomes, turning the action of other genes on and off. She made her discovery in 1951, but it took some 30 years for the significance of her discovery to be recognized!

It has also been found that some sequences of bases along DNA are meaningless. These fragments, called **introns**, are copied into the RNA strand but are edited out before the RNA leaves the nucleus. Introns seem to offer a way of incorporating new information into a gene. The fragments that actually form the gene are called **exons**. Some genes have been found to consist of as many as 52 separate exons.

There is much still to be learned about the structure and function of DNA, but present research indicates that at some stage it will be possible to replace a defective gene in a developing embryo so that it can lead a normal existence. But this raises other questions: What is "normal"? What is "defective"? Who can make such a decision? The significance of such questions makes it important for you, as an informed member of the public, to keep pace with biological research and the associated bioethical issues.

QUESTIONS FOR REVIEW

SOME WORDS TO KNOW

Match each description given in the left-hand column with a word shown in the right-hand column. DO NOT WRITE IN THIS BOOK.

1. Hereditary information found in the nucleus.
2. A nitrogenous base found in DNA.
3. A nitrogenous base found in RNA.
4. The first stage in cell duplication.
5. The process resulting in the formation of RNA.
6. The process resulting in the formation of a protein.
7. The code for the manufacture of a specific protein.
8. The code for selecting a specific amino acid.
9. The portion of tRNA which plugs into a site in mRNA.
10. The enzyme that slices DNA apart.
11. The enzyme that joins DNA together.
12. A virus that infects bacteria.
13. The meaningless portion of DNA.

A. bacteriophage
B. codon
C. intron
D. anticodon
E. guanine
F. DNA ligase
G. DNA
H. DNA replication
I. mRNA
J. uracil
K. translation
L. restriction endonuclease
M. transcription

SOME FACTS TO KNOW

1. What is the chemical composition of a typical cell?
2. Why can there be so many different proteins?
3. Why are the proteins on the surface of the cell so important?
4. Name the four nitrogenous bases found in DNA.
5. What is a nucleotide?
6. How do the nucleotides link together in DNA?
7. Discuss the significance of the complementary structure of DNA.
8. Name the three types of RNA. What is the function of each?
9. How does RNA transcription differ from DNA replication?
10. How does mRNA specify which amino acids belong in each location?
11. What is the significance of the shape of tRNA?
12. What do we now understand a mutation to be?
13. How have bacteriophages aided in DNA research?

QUESTIONS FOR RESEARCH

1. The structure of a number of proteins has been determined. Investigate the techniques used to accomplish this. To help you visualize the complexity of these structures, construct a molecular model of a simple protein.
2. Francis Crick and James Watson received the Nobel prize in 1962 for developing a model for the structure of the DNA molecule. Investigate their research. Construct a molecular model of a segment of DNA.
3. Investigate Barbara McClintock's research on jumping genes. She is still an active researcher. What other contributions has she made?
4. Investigate the organization of the computer network coordinating the availability of organs for transplants. What are the time restrictions under which it must operate?
5. Although their potential is not fully understood, highly specific antibodies called **monoclonal antibodies** are being used in organ transplants and cancer research. Investigate the use of these antibodies to determine their possibilities.
6. It has been suggested that specially tailored viruses carrying the missing gene could be used to infect and cure people with genetic deficiencies. Research the techniques used. Investigate the present concern over potential risks to the non-deficient, "normal" population. What are the risks, if any? Is the concern about them justified, in your opinion?

7. In the early 1970s there was considerable alarm about the new field of recombinant DNA research. Scientists voluntarily stopped doing research in this area until the matter was investigated and guidelines established. Find out how the matter was resolved. What guidelines were instituted to govern such research? How effective are the guidelines? Who should be responsible for regulating such issues: governments? scientists? the public?

8. Find out which of the hereditary diseases might benefit from DNA research. What are the funding sources that support such projects? What research is being done at present?

9. Tiny rings of DNA called **plasmids** are now being used to insert DNA into cells. Investigate the techniques used.

10. Recombinant DNA techniques are also being used to improve agricultural plants. Investigate how such research projects might help to improve the nutrition of the world's hungry people.

Appendices

Appendix 1

Review of Cell Structure and Function

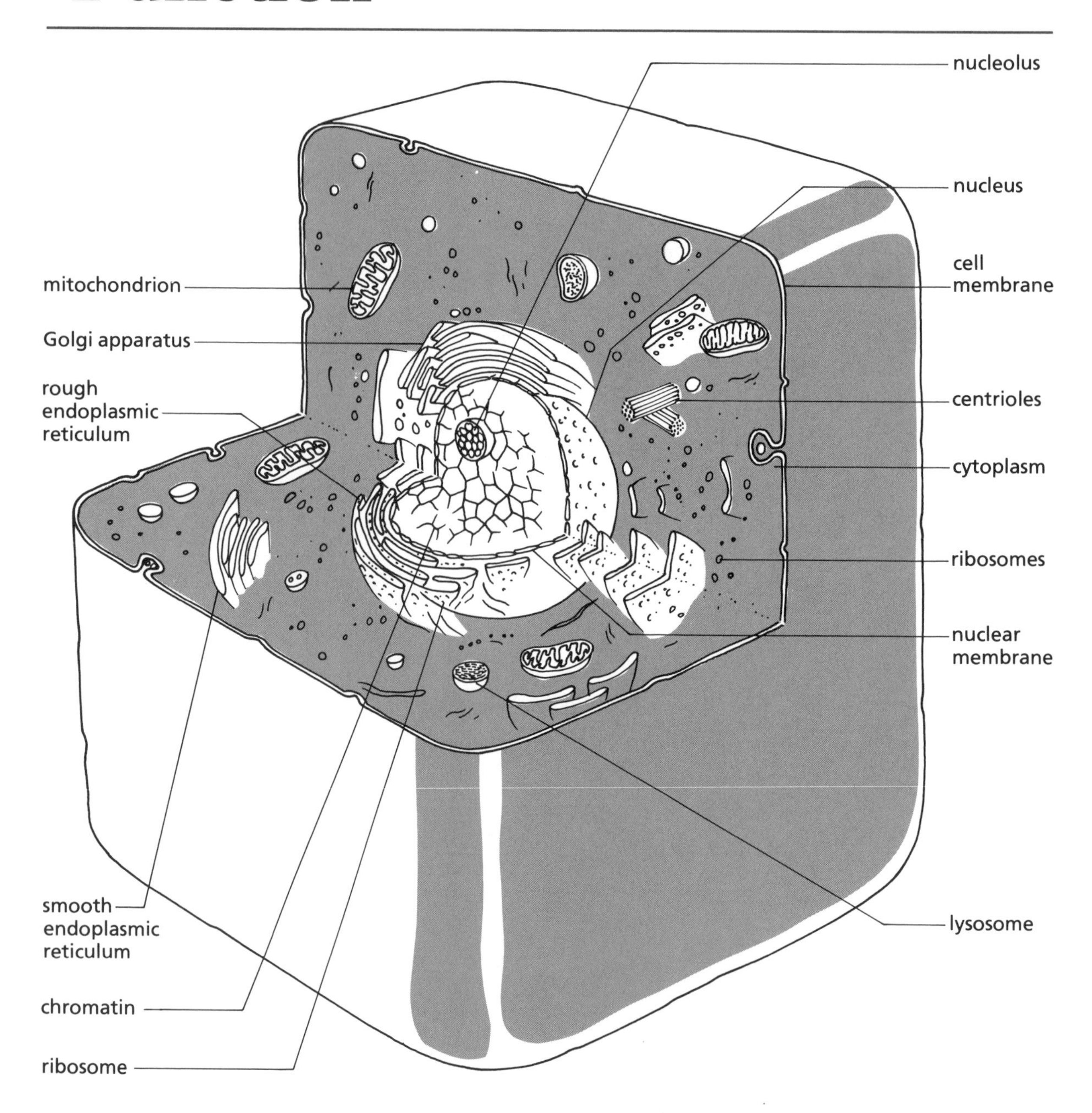

ORGANELLE	FUNCTION
centriole	organizes transfer of DNA during cell duplication
chromatin	network of DNA molecules
rough endoplasmic reticulum	contains ribosomes – stores resulting protein until required
smooth endoplasmic reticulum	site of synthesis of other molecules
Golgi apparatus	delivers enzymes and other proteins to required location
lysosome	container of enzymes
mitochondrion	site of final stages of energy release
nucleus	contains DNA (hereditary information of cell)
nucleolus	site of RNA synthesis
ribosome	site of protein synthesis

Appendix 2

Vitamins For Health

These chemically unrelated organic substances are grouped together because each is essential, in small amounts, in the human diet.

VITAMIN	FUNCTION	DEFICIENCY SYMPTOMS	DAILY REQUIREMENTS (16-18 years of age)	FOOD SOURCE
A	The beauty vitamin. Needed to maintain healthy skin, hair, eyes, etc. Improves resistance to infections. Helps break down fats.	Rough, dry skin, low resistance to infections, night blindness.	girls 4000 IU boys 5000 IU (Toxic if intake is higher.)	Whole milk, liver, butter, carrots, eggs, green and yellow vegatbles.
D	Needed for calcium and phosphorus absorption to produce good bones and teeth. Regulates blood calcium levels.	**Rickets**, softening of bones in adults, and poor teeth.	400 IU (Very toxic if intake is higher.)	Fish liver oils, sardines, salmon, liver. Some made by the skin in sunlight.
E	Helps in the formation of red blood cells, muscle, and other tissues. Prevents abnormal breakdown of fat.	Circulation problems, loss of sexual and body vigour, muscular and heart problems.	girls 7 mg boys 10 mg	Vegetable oils, whole grain cereals.
K	Aids in blood clotting.	Rare, generalized bleeding.	Not required daily; small amounts weekly.	Green vegetables.
Thiamine (B_1)	Needed for oxidation of carbohydrates in Krebs' Cycle.	**Beriberi**, loss of energy, depression, poor appetite, skin problems.	girls 1.1 mg boys 1.5 mg	Whole grain cereals, dry yeast, pork, fish, lean meat.

Riboflavin (B_2)	Used as electron carrier in respiratory chain formation of ATP. Also involved in metabolism of fatty acids.	Tissue damage, eye strain, fatigue, itching, sensitivity to light.	girls 1.4 mg boys 1.8 mg	Liver, milk, cheese, green leafy vegetables, beans.
Niacin (B_3)	Involved in energy release reactions in respiratory chain. Also involved in fatty acid synthesis.	**Pellegra**, lack of concentration, insomnia, headaches, backache, poor memory.	girls 14 mg boys 20 mg	Meat, poultry, fish, whole wheat and enriched grains.
B_{12}	Important in DNA synthesis. Essential for proper functioning of the nervous system.	**Anemia**, bowel disorders, poor appetite, and poor growth.	1.9 μg	Liver, kidney, fish.
Ascorbic Acid (C)	Helps to maintain normal development of bones, teeth, gums, and cartilage.	**Scurvy**, bleeding gums, easy bruising, low resistance to infections.	girls 45 mg boys 55 mg Smokers need more.	Citrus fruits, green vegetables, potatoes.

Glossary

A

acclimation. The long-term adjustment of metabolic activity to environmental change.

acid. A solution that contains H_3O^+ ions. Also, any substance that gives off H^+ ions in a chemical reaction.

activation energy. The energy required to initiate a chemical reaction.

ADP **(adenosine diphosphate).** A substance used to store energy for cell functions when one phosphate group is added to form ATP.

aerobic. Occurring in the presence of oxygen.

amino acid. An organic acid containing the NH_2 group. Amino acids are the building blocks of proteins.

anabolism. The synthesis of larger molecules from smaller ones.

anaerobic. Occurring in the absence of oxygen.

anticodon. The portion of the tRNA molecule that fits into the codon of the mRNA during protein synthesis on the ribosome.

atom. The smallest particle of an element that still possesses the properties of that substance.

ATP **(adenosine triphosphate).** A substance that is the primary energy source for cells because of the energy released when one phosphate group is removed from it to form adenosine diphosphate (ADP).

B

bacteriophage. A virus that infects bacteria.

base. A substance that produces OH^- ions in a water solution. Also, any substance that gains H^+ ions in a chemical reaction.

biotechnology. The industrial application of biological processes.

bond energy. The amount of energy released during the formation of a chemical bond.

bonding sites. Locations at which atoms can readily join together or split apart during chemical reactions.

buffer. A solution that can maintain a stable pH within certain limits of exposure to acids or bases.

C

carbohydrate. A molecule containing carbon, hydrogen, and oxygen, in which the ratio of hydrogen atoms to carbon atoms is two to one.

catabolism. The breaking apart of larger molecules into smaller ones.

catalyst. A substance that decreases the activation energy of a reaction.

cellular respiration. The process through which the cell releases energy from glucose.

chromosome. A structure within the nucleus, composed of DNA, that transfers hereditary characteristics from a cell to its offspring.

codon. A sequence of three bases that specifies a particular amino acid; found on the mRNA molecule.

co-enzyme. A substance (often a vitamin) that forms a complex which activates an enzyme.

co-factor. A metallic ion needed to activate an enzyme.

competitive inhibition. The blocking of the action of an enzyme by an incorrect substrate.

compound. A substance containing two or more kinds of atoms united in a definite proportion. The properties of a compound differ from those of the component elements.

control centre. The portion of a feedback system that selects the appropriate response.

covalent bond. A chemical bond formed by the sharing of electrons between atoms.

D

dialysis. A process involving the movement of molecules across a differentially permeable membrane.

digestion. The breakdown of complex carbohydrates, proteins, and fats into absorbable units.

dipeptide. A molecule consisting of two linked amino acids.

DNA **(deoxyribonucleic acid).** A complex molecule that codes the information which enables a cell to manufacture proteins.

DNA **ligase.** An enzyme involved in the repair of DNA that causes the DNA fragments to adhere to each other.

DNA **polymerases.** Enzymes involved in the synthesis or repair of DNA that cause base pairs to join.

dehydration. The removal of water.

disaccharide. A molecule consisting of two linked monosaccharides.

E

electron. A tiny, negatively charged particle found moving around the nuclei of atoms.

electronegativity. The ability of an atom to attract electrons in an ionic bond. Electronegativity provides a measure of the tendency of an element to participate in ionic bonding.

element. A substance containing only one kind of atom.

endocrine glands. Glands that secrete their products into the blood.

endothermic reaction. A reaction that results in a net intake of energy.

enzyme. A protein that catalyzes a specific biological reaction.

enzyme repression. A process in which the product of a reaction prevents the formation of the enzyme which catalyzes that reaction.

exocrine glands. Glands that secrete their products outside the body or in the digestive cavity.

exon. A meaningful sequence of bases in the DNA coding for the manufacture of a specific protein.

exothermic reaction. A reaction that results in a net release of energy.

F

fatty acid. A long chain organic acid forming part of a lipid molecule.

feedback inhibition. A process in which the product of a reaction inhibits the first enzyme facilitating it.

feedback loop. A system capable of adjusting its response to correspond to changes in conditions.

fermentation. A series of chemical reactions that release a small amount of energy from pyruvic acid in the absence of oxygen.

fluids. Substances that flow easily.

G

gene. The portion of the DNA molecule responsible for the formation of a specific protein.

glycolysis. The initial series of chemical reactions in cellular respiration that release energy from glucose.

H

homeostatic mechanism. A feedback system that helps an organism maintain a stable internal environment.

hormone. A chemical released by endocrine glands into the bloodstream to produce a response in a target organ.

hydrocarbon. A substance that contains only carbon and hydrogen.

hydrolysis. A process of splitting molecules that involves the addition of a water molecule.

hypothermia. A severe lowering of the body's core temperature.

I

inhibit. To repress or stop.

intron. An apparently meaningless segment in the sequence of bases in DNA.

ionic bonds. Chemical bonds formed by the mutual attraction of oppositely charged ions.

ions. Atoms or groups of atoms that have become electrically charged by the gain or loss of electrons.

Islets of Langerhans. Clusters of cells found in the pancreas.

isomers. A group of different substances that contain the same numbers and kinds of atoms linked in different arrangements.

K

Kreb's Cycle. A cyclic series of chemical reactions that release energy from pyruvic acid in the presence of oxygen.

L

lipid. A fat or oil.

M

medulla oblongata. The central portion of the brain; controls body functions essential to life.

metabolic rate. The speed with which cells carry out their chemical activities.

mitochondrion. A cell organelle containing internal membranes along which are arranged the enzymes required in Kreb's Cycle and the respiratory chain.

mole. A standard number (6.02×10^{23}) used in expressing quantities of substances.

monitor. The portion of a feedback system that detects change.

monosaccharides. Single sugars.

mutation. An alteration in the sequence of bases in the DNA molecule.

N

neutron. A small, uncharged particle found in the nuclei of atoms.

nitrogenous base. Any of five nitrogen-containing molecules that are components of DNA or RNA: adenine, guanine, thymine, cytosine, or uracil.

nucleotide. A segment of the DNA molecule consisting of a phosphate group, a sugar, and a nitrogenous base.

O

organelle. A membrane-surrounded area within a cell; specialized to perform a specific function.

oxygen debt mechanism. The requirement for oxygen to metabolize the lactic acid accumulated during fermentation.

P

peptidase. An enzyme that breaks a peptide bond.

peptide bond. A chemical bond that forms between a carbon atom of one amino acid and a nitrogen atom of another.

pesticide. A substance designed to kill an organism that interferes with human activities.

pH indicator. A substance whose molecular structure, and therefore colour, changes in response to changes in pH.

pH scale. A scale comparing the H_3O^+ ion concentration (acidity) of solutions.

polar molecule. A molecule containing bonds in which the electrons are unequally shared between atoms.

polymer. A large molecule consisting of one or more chains of simpler molecules.

polysaccharides. Molecules that consist of long chains of monosaccharides.

protein. A complex molecule composed of many amino acids.

proton. A small, positively charged particle found in the nuclei of atoms.

R

receptor. A group of cells that respond to a specific stimulus.

recombinant DNA. A DNA molecule containing an inserted nucleotide fragment.

regulator. The portion of a feedback system that makes the necessary adjustments.

replication. The formation of a duplicate strand of DNA using the complementary strand as a template.

respiratory chain. The series of reactions that use the energy obtained during glycolysis and Kreb's Cycle to form ATP.

restriction endonucleases. Enzymes that sever DNA between specific sequences of bases. These enzymes are used extensively in recombinant DNA research.

reverse transcriptase. An enzyme causing the synthesis of a DNA fragment complementary to RNA.

Rh factor. A protein found on the surface of blood cells. It is one of the determinants of blood type.

ribosome. A tiny cell organelle required for protein synthesis.

RNA (ribonucleic acid). A molecule that transfers the coded information for a specific protein from DNA to a ribosome.

S

saturated compound. A substance containing only single bonds between atoms.

synthesis. The process through which large molecules are formed from smaller ones.

structural formula. A chemical formula that shows the arrangement of atoms within a molecule.

substrate. The substance acted upon by an enzyme.

T

transcription. The formation of a strand of RNA using a segment of DNA as a template.

translation. The assembly of amino acids to form a protein in the order specified by RNA.

U

unsaturated compound. A substance containing double or triple bonds.

Index

Note: When a page number is preceded by an "f", the reference is to a figure on that page.

A

B

C

H

I

J

K

L

M

N

O

P

Credits

Photographs and Illustrations:

Title page: Woman looking into microscope: Health Sciences Media Services, Sunnybrook Medical Centre

Chapter 1 Opener: Bicycle racers: The Ontario Cycling Association; p. 4: Model of electrons: Birgitte Nielsen

Chapter 2 Opener: Girls on ride: National Film Board of Canada; Figure 2.8: Greg Lockhart

Chapter 3 Opener: Boy doing handstands: Tomislov Zivic and Langstaff Gymnastic Club; Figure 3.5: The Canadian and Ontario Cycling Associations; Figure 3.8: Women's College Hospital, Toronto; Figure 3.13: Ontario Pesticides Advisory Committee and Robert McDonald

Chapter 4 Opener: Children on jungle gym: Francis Roque; Figures 4.8, 4.9, 4.10, and 4.11: Adapted from Gideon E. Nelson, *Biological Principles with Human Perspectives*, 2nd edition (New York: John Wiley & Sons, Inc., 1980, 1984), p. 263; Figure 4.12: The Hospital for Sick Children, Toronto